Advances in Oil and Gas Exploration & Production

The book series *Advances in Oil and Gas Exploration & Production* is dedicated to publishing scientific monographs that cover a broad range of topics in geophysical and geological research related to both conventional and unconventional oil and gas systems. These topics are explored from both exploration and production perspectives.

The series aims to build a comprehensive and diverse library of reference works that reflect the current state of research on selected themes—ranging from specific techniques employed in petroleum geoscience to regionally focused studies. All volumes are authored and edited by leading experts actively working in their respective fields.

The *Advances in Oil and Gas Exploration & Production* series includes single- and multi-authored books, edited volumes, and conference proceedings. A Book Proposal Form is available on our website, or you may request one directly from the Publishing Editor, Qiao Shu, at **qiao.shu@springernature.com**.

Carlos Urdaneta · Aamir Bader Shah ·
Xuqing Wu · Xin Fu · Jiefu Chen

Deep Learning Applications for Drilling Performance: Forecasting, Diagnostics, Telemetry, and Predictive Maintenance

Springer

Carlos Urdaneta
Department of Electrical
and Computer Engineering
University of Houston
Houston, TX, USA

Aamir Bader Shah
Department of Electrical
and Computer Engineering
University of Houston
Houston, TX, USA

Xuqing Wu
Department of Information Science
Technology
University of Houston
Houston, TX, USA

Xin Fu
Department of Electrical
and Computer Engineering
University of Houston
Houston, TX, USA

Jiefu Chen
Department of Electrical
and Computer Engineering
University of Houston
Houston, TX, USA

ISSN 2509-372X ISSN 2509-3738 (electronic)
Advances in Oil and Gas Exploration & Production
ISBN 978-3-032-26647-7 ISBN 978-3-032-26648-4 (eBook)
https://doi.org/10.1007/978-3-032-26648-4

This Springer imprint is published by the registered company Springer Nature Switzerland AG
The registered company address is: Gewerbestrasse 11, 6330 Cham, Switzerland

Competing Interests The authors have no competing interests to declare that are relevant to the content of this manuscript.

Contents

Introduction and Background 1

1.1 Background

The global shift toward low-carbon energy systems has amplified interest in technologies such as geothermal energy and carbon capture and storage (CCS). These solutions offer promising pathways to reduce greenhouse gas emissions, yet their deployment remains limited by operational, financial, and technical barriers. Central to both geothermal and CCS development are drilling operations, which represent one of the most cost-intensive and risk-prone components of project execution. Improving drilling reliability and efficiency is therefore critical not only for cost reduction but also for accelerating the adoption of sustainable subsurface technologies.

The drilling industry, including geothermal, CCS, and conventional oil and gas operations, continues to face enduring performance challenges. Operators must balance competing objectives such as minimizing non-productive time (NPT), optimizing rate of penetration (ROP), preserving tool integrity, and managing project costs. These difficulties are magnified in technically demanding scenarios such as high-pressure and high-temperature (HPHT) wells, extended-reach drilling (ERD), underbalanced drilling (UBD), and operations conducted in geologically complex formations. Such conditions introduce high temperature, shock, and vibration, all of which degrade tool performance and operational predictability.

Compounding these technical challenges is the variability of subsurface conditions encountered during drilling. Formation transitions and downhole equipment wear introduce nonlinearities that disrupt the expected relationships between control parameters and drilling outcomes. Historically, these issues have been approached using traditional optimization strategies centered on physics-based models, offset well analysis, and pre-drill simulations. While effective for initial planning, these methods rely on fixed assumptions that limit their ability to adapt in real time. Manual intervention by directional drillers and operations engineers remains necessary to identify anomalies during operations. This reactive approach increases the risk of misinformed decisions and operational downtime.

A key limitation of conventional drilling optimization lies in its inability to fully model the dynamic coupling between control inputs and mechanical response. Variables such as weight on bit (WOB), revolutions per minute (RPM), pump rate, and mud properties interact in ways that are highly dependent on formation-specific characteristics and tool configuration. Existing models often assume linear or time-invariant behavior, which oversimplifies the complexity of real-world drilling conditions. As a result, optimization workflows based on these models frequently underperform in the field, leading to suboptimal ROP and accelerated equipment wear.

Artificial intelligence (AI) offers new opportunities to improve drilling workflows

C. Urdaneta et al., *Deep Learning Applications for Drilling Performance: Forecasting, Diagnostics, Telemetry, and Predictive Maintenance*, Advances in Oil and Gas Exploration & Production,
https://doi.org/10.1007/978-3-032-26648-4_1

through real-time pattern recognition, predictive modeling, and adaptive control based on live sensor inputs. This book investigates transformer, deep autoencoder, and variational autoencoder architectures that support forecasting of drilling dynamics, early detection of tool degradation, and denoising of telemetry signals. These models identify evolving patterns in downhole and surface measurements. They also enhance the clarity of electromagnetic telemetry signals by suppressing environmental noise, which is especially important in high noise environments or high-attenuation formations. By improving drilling dynamics prediction accuracy, tool health monitoring, and telemetry reliability, these approaches enable more reliable operational decisions, help prevent NPT, and extend the service life of downhole tools.

Prediction and optimization of ROP is one of the most practical applications of AI in drilling. Traditional ROP models typically use empirical correlations and regression equations that assume fixed relationships between control variables, making them less effective in capturing the complex factors present in drilling operations. In contrast, the time-series learning algorithms employed in this research continuously refine their predictions based on real-time inputs such as WOB, RPM, torque, pump rate, and pressure. These models are capable of capturing nonlinear interactions among drilling parameters, which is essential for accurate forecasting under variable conditions. Because ROP directly impacts drilling cost, tool wear, and carbon footprint, achieving optimal ROP depends on careful adjustment of multiple parameters in relation to formation properties. AI models that learn from both historical and real-time data offer a more adaptive approach, improving performance in dynamic formations where conventional models fall short.

In addition to optimizing ROP, AI models contribute to the detection and mitigation of common drilling dysfunctions such as stick–slip, bit balling, and lateral shock. These conditions interfere with predictable bit-rock interaction and introduce erratic tool behavior that degrades drilling performance. As shown in Fig. 1.1, such dysfunctions accelerate bit wear and increase the likelihood of tool failure. Continuous monitoring and early detection of these dysfunctions are essential for minimizing risk, but traditional control systems are limited in their ability to adapt quickly. AI-based diagnostics improve situational awareness by learning the characteristic patterns of instability and providing early indicators for corrective action.

Fig. 1.1 Accelerated bit wear (Rickard et al. 2019)

Coiled tubing drilling (CTD) has emerged as an efficient alternative to conventional rotary drilling, particularly in complex wellbore geometries and environments requiring precise pressure control. CTD systems are compact, mobile, and capable of underbalanced drilling, minimizing formation damage and enhancing reservoir productivity (Fig. 1.2). However, CTD operations are highly sensitive to downhole vibrations and shock dynamics, which pose operational challenges. Effective prediction of these dynamics are essential to prevent NPT and equipment failure while maintaining drilling efficiency.

Modern drilling workflows are increasingly supported by measurement while drilling (MWD) tools, which utilize advanced sensors to provide real-time data on critical parameters such as WOB, RPM, pressure, temperature, and vibration (Fig. 1.3). These tools play an essential role in optimizing drilling performance, detecting anomalies, and enabling

Fig. 1.2 CTD rig (Urdaneta et al. 2024)

dynamic adjustments. However, MWD data often suffer from noise, inconsistencies, and occasional sensor failures, limiting the reliability of real-time insights.

The availability of high-resolution time-series data from MWD tools and other downhole sensors has created new opportunities for improving drilling performance through data-driven modeling. These data streams capture dynamic interactions between geological conditions, tool behavior, and operational adjustments in real time, making them well-suited for AI methods that learn from sequential patterns. This book focuses on transformer, deep autoencoder, and variational autoencoder models, each applied to distinct challenges in modern drilling workflows, including forecasting ROP and shock, diagnosing tool health anomalies, and improving the reliability of electromagnetic telemetry under noisy conditions.

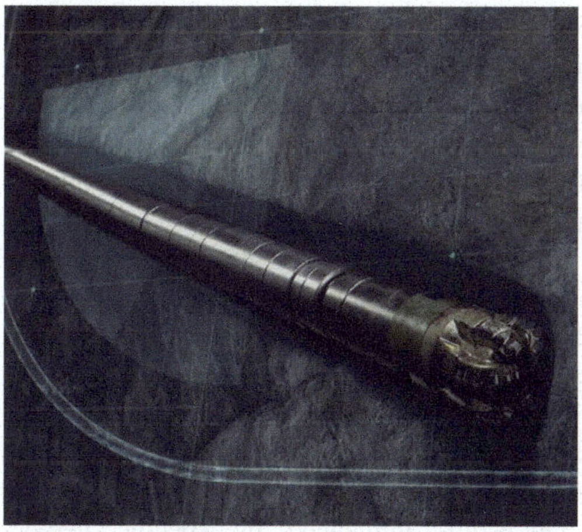

Fig. 1.3 MWD tool (Urdaneta et al. 2024)

Transformer architectures are used for forecasting ROP and downhole shock, where long-range dependencies across operational inputs are critical to capturing temporal shifts in drilling response. Compared to earlier neural network approaches, transformers offer more consistent performance across variable drilling conditions by learning from both short-term transitions and broader operational trends. For health diagnostics, variational autoencoders are applied to resistivity tool measurements to detect anomalous signal behavior without requiring labeled fault data. This enables early identification of tool degradation that might otherwise go unnoticed in traditional monitoring systems. For telemetry improvement, deep autoencoders with multi-layer encoder-decoder structures are trained to recover clean signal profiles from noisy electromagnetic transmissions, enhancing downhole communication reliability in environments prone to interference.

Because predictions affect safety–critical decisions, model outputs must be transparent and interpretable to operators. To support this, a confidence scoring system is introduced, combining multiple uncertainty estimation methods in a multi-expert ensemble to quantify prediction reliability. This enables the system to flag low-confidence forecasts, providing additional context during operational decision-making and reinforcing trust in AI-assisted recommendations.

Together, these models represent the foundation for the investigations that follow in this book. Chapters 2 through 8 expand on each technical element: Chap. 2 reviews the state of AI-driven drilling technologies; Chaps. 3 to 6 demonstrate forecasting, diagnostics, and telemetry-improvement methods using transformer and autoencoder architectures; Chap. 7 extends these concepts to predictive maintenance; and Chap. 8 explores synthetic data and hybrid modeling strategies that enhance deep-learning generalization. Chapter 9 examines cross-cutting limitations, governance challenges, and future directions for deploying these systems in field operations.

1.2 Problem Statement

Efficient drilling operations are central to reducing operational costs, facilitating the broader deployment of sustainable subsurface technologies, including geothermal energy and CCS. Despite continuous advancements in drilling modeling techniques, achieving consistently optimal drilling performance remains challenging. These complexities come from the dynamic and nonlinear characteristics in drilling processes. Critical performance indicators such as ROP and downhole shock dynamics are particularly influential as they directly impact operational efficiency. However, accurate real-time prediction of these critical metrics remains elusive due to ongoing limitations in conventional modeling methods.

Traditional approaches to drilling optimization rely on empirical relationships, regression-based predictive equations, and the accumulated expertise of drilling operators. Central to these traditional methods is the concept of mechanical specific energy (MSE), which quantifies the energy required to remove a unit volume of rock. According to Rickard et al. (2019), MSE incorporates two fundamental components: rotational energy and axial energy. The rotational energy component reflects the work done by torque and rotational speed, incorporating the bit diameter and ROP. The axial component, on the other hand, considers the weight applied on the drill bit and its corresponding cross-sectional area. In the formulation used by Rickard et al. (2019), MSE is expressed as

$$MSE = \frac{480\,T\,RPM}{d_{bit}^2\,ROP} + \frac{4\,WOB}{\pi\,d_{bit}^2} \tag{1.1}$$

where T is torque, RPM is rotational speed, d_{bit} is bit diameter, ROP is rate of penetration, and WOB is weight on bit. Collectively, these terms provide essential insights into how effectively the input mechanical energy translates into rock penetration. In practical scenarios, lower MSE values indicate efficient drilling conditions, whereas elevated values typically signal inefficiencies arising from factors such as excessive bit wear, suboptimal operational parameters, or persistent drilling dysfunctions.

Despite its widespread adoption, empirical MSE models are often insufficiently responsive to accurately represent the intricate and nonlinear interactions between drilling parameters under dynamically changing geological environments. Real-time adaptation and effective management of these interactions require advanced predictive models capable of continuously adjusting to evolving operational contexts, a capability notably absent in traditional modeling paradigms.

Further complicating the effectiveness of conventional drilling optimization methods is their general inability to comprehensively capture the intricate interplay among key operational parameters, including WOB, RPM, torque, pressure, and fluid flow rate. These parameters frequently exhibit complex and nonlinear relationships influenced by specific geological formations and varied operational contexts.

Traditional models, which typically assume linear or static interactions among these variables, are often inadequate when attempting to predict drilling performance under real-world operational conditions. Additionally, conventional approaches regularly overlook vital yet difficult-to-measure parameters, such as the pressure gradient ahead of the drill bit, the compressibility characteristics of the formation, and progressive wear patterns of drilling tools. Each of these neglected factors exerts considerable influence on drilling performance, yet traditional optimization strategies rarely account for them, thereby diminishing the accuracy and practical applicability of model predictions.

Recent advancements in AI and machine learning (ML) present promising tools capable of addressing the limitations in traditional drilling optimization methods. The transformer architecture has emerged as a particularly effective class of ML models because it can process and analyze sequential data. Originally developed within the context of natural language processing, transformers excel in recognizing and modeling both short-term fluctuations and long-term dependencies within time-series data. Transformers demonstrate considerable potential to effectively address and model the nonlinear interactions among drilling parameters, operational conditions, and mechanical responses of drilling equipment.

However, despite their potential, several challenges remain associated with implementing AI models within drilling optimization frameworks. A primary issue lies in the quality of drilling datasets, particularly those obtained from MWD tools. Such datasets frequently include essential parameters like WOB, torque, RPM, and downhole shock measurements but are typically characterized by incomplete records, sensor noise, and variability across different wells and geological formations. Sensor malfunctions, irregular sampling intervals, and environmental disturbances exacerbate these data quality issues, undermining the accuracy of predictive models developed from these datasets.

Additionally, interpretability remains a barrier to operational adoption of AI models within drilling environments. Drilling operators routinely confront high-risk decisions informed by model outputs. Consequently, operators must understand the reliability of these model outputs. However, current AI models often lack indicators to help operators interpret model predictions. Without mechanisms to quantify prediction uncertainty, operator trust in AI-generated insights diminish. Establishing reliable confidence scoring mechanisms is important for informed decision-making and operational trust in AI-based models.

1.3 Research Objectives

This book aims to advance AI applications in drilling operations by focusing on several research objectives, each corresponding to specific chapters within the book. These objectives were selected to address persistent challenges in geothermal, coiled-tubing, and conventional drilling environments while establishing a unified data-driven framework for drilling optimization.

The first research objective is to develop and evaluate transformer-based architectures tailored for predicting drilling dynamics such as rate of penetration (ROP) and downhole shock. Chapters 3 and 4 explore these applications separately for geothermal and coiled-tubing operations. Traditional models, including empirical correlations and recurrent networks, often lose accuracy under variable lithology or mechanical loading. Transformer architectures are introduced as an alternative capable of modeling both short- and long-term temporal dependencies. The geothermal studies in Chap. 3 draw on datasets from the Utah FORGE project, while Chap. 4 demonstrates adaptability within coiled-tubing drilling, characterized by higher vibration frequencies and nonlinear tool responses.

The second objective is to develop diagnostic methodologies for downhole tool health assessment. Chapter 5 introduces a variational autoencoder-based framework for detecting resistivity tool anomalies. Resistivity measurements are critical for geosteering and formation evaluation but are prone to gradual degradation that may precede sensor failure. By learning nominal operational patterns and quantifying reconstruction errors, the model can identify early deviations from expected behavior, improving reliability in directional drilling.

The third research objective extends deep-learning applications to electromagnetic telemetry. Chapter 6 examines deep autoencoder architectures designed to enhance signal quality by filtering noise and reconstructing corrupted signals. Because telemetry systems frequently operate in harsh or high-resistivity formations, the chapter emphasizes real-time denoising and the preservation of meaningful signal content. These methods directly improve the dependability of surface-to-downhole communication during automated drilling operations.

The fourth objective introduces predictive maintenance for drilling equipment. Chapter 7 focuses on remaining-useful-life (RUL) prediction and probabilistic modeling of tool degradation using machine-learning and statistical frameworks. By combining operational logs, vibration data, and prior failure records, these models estimate component longevity and optimize replacement schedules, aligning with proactive asset-management strategies.

The fifth objective explores hybrid and synthetic data generation methods to strengthen model generalization. Chapter 8 presents a framework for combining physics-guided neural architectures with synthetic data produced through augmentation and simulation. These techniques expand the training space available for drilling-related forecasting tasks and enhance model robustness in low-label or rare-event scenarios.

1.4 Organization of the Book

This book is organized to provide a progression from foundational concepts to advanced implementations and future perspectives. Each chapter focuses on a specific domain where AI provides measurable value to drilling performance.

Chapter 2 reviews the state of the art in AI-driven drilling technologies. It opens with a focused theoretical primer that defines core artificial intelligence concepts used throughout the book, clarifies what these models can and cannot infer from drilling data, and summarizes practical interpretability expectations for engineering decision support. Then, it surveys recent literature on digital transformation, data workflows, automation, and real-time pattern recognition, defining the digital and organizational context within which the following research was conducted.

Chapter 3 focuses on rate-of-penetration forecasting for geothermal drilling. It applies transformer architectures to high-temperature datasets from the Utah FORGE project, addressing challenges related to extreme thermal gradients, formation variability, and sensor reliability. Results demonstrate how temporal-attention models outperform conventional networks in predicting ROP under dynamic subsurface conditions.

Chapter 4 extends transformer-based forecasting to coiled-tubing drilling operations. The chapter investigates the prediction of ROP and downhole shock to support vibration mitigation and control. By modeling nonlinear tool responses and frequent transient shocks, the transformer framework provides early warnings for mechanical stress and potential tool wear.

Chapter 5 introduces an unsupervised learning framework for resistivity tool diagnostics. Using variational autoencoders, the chapter demonstrates anomaly detection in resistivity measurements, enabling early identification of tool degradation and supporting reliable formation evaluation and geosteering.

Chapter 6 presents deep-autoencoder-based techniques for enhancing electromagnetic telemetry reliability. It analyzes noise suppression, signal reconstruction, and low-bandwidth communication in deep or high-resistivity wells, showcasing AI-driven denoising methods that improve real-time data integrity.

Chapter 7 examines predictive maintenance and remaining-useful-life modeling for drilling equipment. This chapter integrates sensor data, operational logs, and environmental parameters to forecast tool degradation patterns, contributing to risk reduction and cost efficiency in long-term drilling campaigns.

Chapter 8 explores synthetic data and hybrid AI modeling for deep-learning generalization. It combines physics-guided and data-driven approaches, leveraging simulation, augmentation, and domain transfer to address data scarcity and enhance the robustness of drilling-forecasting models.

Chapter 9 addresses challenges, limitations, and future research directions. It synthesizes key cross-cutting issues, including data integrity, uncertainty quantification, model interpretability, ethical considerations, and industrial deployment barriers, outlining pathways for continued advancement.

References

Rickard, W. M., McLennan, J., Islam, N., & Rivas, E. (2019). Mechanical specific energy analysis of the FORGE Utah well. In Proceedings of the 44th Workshop on Geothermal Reservoir Engineering. Stanford University.

Urdaneta, C., Jeong, C., Wu, X., & Chen, J. (2024). Deep learning method for improving rate of penetration prediction in drilling. SPE Journal.

State of the Art in AI-Driven Drilling Technologies

2

2.1 Theoretical Introduction for Artificial Intelligence in Drilling Engineering

The chapters in this book emphasize applied deep learning systems for forecasting, diagnostics, telemetry enhancement, and predictive maintenance. Because these models are increasingly used to support operational decisions, a concise theoretical framing is useful to clarify what the models learn, what they do not learn, and how their outputs should be interpreted in an engineering workflow. This section defines the core learning setups used in the book, outlines inherent limitations that commonly appear in drilling datasets, and summarizes interpretability expectations that align with safety and reliability requirements in field operations.

2.1.1 What Artificial Intelligence Models Learn

Most artificial intelligence methods used for drilling applications can be described as learning a mapping from inputs to outputs using historical examples. In supervised learning, the training set contains input signals and a target label or value, and the model learns parameters that minimize a loss function measuring disagreement between predictions and observed targets. Forecasting rate of penetration and shock dynamics fall into this category, where the target is a future value of a drilling metric and the inputs are time indexed sensor measurements and operational parameters.

In unsupervised learning, there is no explicit target label. The model learns a representation of normal behavior, then flags deviations as abnormal. Autoencoders and variational autoencoders are examples used for diagnostics and denoising settings. In a diagnostic workflow, the model is trained to reconstruct nominal sensor patterns. At inference time, unusually large reconstruction error or an atypical latent representation can indicate that the current behavior does not resemble the training distribution.

A third family, generative modeling, learns an internal distribution over plausible data patterns. Variational autoencoders are generative in this sense because they learn a latent probability model that can produce samples resembling the training data. In drilling contexts, this is often used to address label scarcity by expanding coverage of nominal patterns, exploring plausible perturbations, or supporting stress tests of anomaly thresholds, while recognizing that generated samples are still constrained by the assumptions and coverage of the original dataset.

Across these learning setups, the central point is that the models do not discover first principles physics by default. They learn statistical regularities present in the measured data, shaped by sensor placement, preprocessing choices, and operational practices. As a result, model performance and credibility must be evaluated with

C. Urdaneta et al., *Deep Learning Applications for Drilling Performance: Forecasting, Diagnostics, Telemetry, and Predictive Maintenance*, Advances in Oil and Gas Exploration & Production,
https://doi.org/10.1007/978-3-032-26648-4_2

respect to the engineering decision the model is intended to support.

2.1.2 Time Series Structure in Transformers

Drilling data are sequential and strongly dependent on operational regime. A forecasting model must use the recent past to infer the near future, while accounting for both short timescale transients and longer timescale trends. Traditional sequence models such as recurrent networks compress past information through a hidden state, which can limit how well they retain long range dependencies. Transformer models address this by using attention mechanisms that allow the model to reference different parts of the input sequence when producing predictions. In engineering terms, attention can be viewed as a learned weighting over historical context that helps the model decide which prior conditions are relevant to the current prediction.

However, attention does not guarantee interpretability on its own, and it does not guarantee generalization. A transformer can still overfit to spurious correlations, memorize a narrow operating envelope, or degrade when sensors drift. It can also produce confident predictions in regions where it has little training support unless explicit uncertainty estimation and validation checks are included.

2.1.3 Limitations in Drilling Applications of Artificial Intelligence

Several limitations recur across drilling environments and are important for evaluating any applied study.

Data quality and sensor constraints. Measurement while drilling and surface sensor streams are affected by noise, dropouts, calibration differences, and changes in sampling rate. Preprocessing can partially correct these issues, but it can also introduce unintended artifacts if filters, imputation, or resampling change the statistical structure of the signal.

Non stationarity and regime shifts. Drilling behavior changes across formations, bit wear states, hydraulics settings, and operational phases such as connections or reaming. A model trained on one subset of regimes may perform poorly when the operational state changes. Even within a single well, the data distribution can shift with depth and time.

Limited labels and delayed ground truth. Many high impact events are rare, and failure labels may be incomplete, inconsistently recorded, or only observable after the fact. This limits purely supervised approaches and motivates unsupervised diagnostics and synthetic data strategies, but it also increases the risk of biased evaluation if the test set does not reflect the full space of operational variability.

Correlation does not imply causation. Models trained on observational drilling data can exploit correlations created by operational policy rather than physical causation. For example, if a certain parameter combination is only used in specific lithologies or by specific crews, the model may learn crew or policy signatures rather than rock response. This matters when the model is deployed in a new field or under a different drilling program.

Distribution shift and generalizability. The most common failure mode in deployment is that the model encounters conditions outside the training distribution. Generalizability should therefore be treated as a primary engineering requirement rather than a secondary performance detail. Evaluation should include tests across wells, sections, and operational regimes whenever feasible, and results should be framed as conditional on the dataset coverage.

Objective mismatch and metric selection. Training losses often differ from operational metrics. For probabilistic forecasting, likelihood based losses encourage calibrated predictive distributions, while point error metrics emphasize average accuracy. If a model is used for decision support, the training objective should be aligned with the decision risk, for example, emphasizing robustness to outliers or penalizing large errors more heavily when they lead to unsafe actions.

Compute and latency constraints. Real time decision support requires inference within tight time windows and within the computing constraints of rig or edge systems. Larger models may be accurate offline but impractical in deployment unless compressed, distilled, or engineered for efficient inference.

These limitations define the operating conditions under which model results should be trusted, and the checks needed when conditions change. They motivate careful validation, uncertainty quantification, and interpretability practices aligned with the decision context.

2.1.4 Interpretability in an Engineering Context

Interpretability can mean different things depending on the decision being supported. In drilling operations, interpretability should be defined as the ability for a practitioner to understand, at an appropriate level of abstraction, why the model produced a recommendation or prediction, how reliable it is likely to be, and what conditions might cause it to fail. This framing aligns with the operational trust barrier already identified in the problem statement, where operators must understand reliability rather than treat predictions as unquestionable outputs.

Practical interpretability can be achieved through a combination of mechanisms.

Feature attribution and sensitivity. Post hoc methods can estimate which input channels most influenced a prediction over a given time window. In drilling, this can be presented as relative influence of parameters such as weight on bit, rotational speed, torque, pressure, and recent vibration statistics, with the key caveat that attribution reflects association within the model, not physical causality.

Example based explanations. For time series models, a useful engineering explanation is often analogical: identifying similar historical patterns or operating windows in the training set that resemble current conditions. This supports operator intuition and can reveal when the model is extrapolating.

Uncertainty and confidence scoring. For safety relevant workflows, interpretability should include a reliability signal. A prediction without a confidence estimate forces an operator to guess whether the model is operating inside or outside its competence region. Confidence scoring and uncertainty quantification methods provide an operationally meaningful layer of interpretability by separating high confidence predictions from cases where human review or fallback logic is appropriate. This is consistent with the book's emphasis on quantifying reliability of outputs to reinforce trust in decision making.

Constraints and plausibility checks. Engineering interpretability is strengthened when model outputs are filtered through basic physical plausibility checks, bounds, and consistency rules. These checks do not replace learning, but they reduce the risk of extreme recommendations caused by sensor glitches or out of distribution inputs.

Human in the loop deployment. Interpretability is ultimately validated in use. The goal is not to eliminate human judgment but to provide decision support that is transparent about confidence, failure modes, and data requirements. This perspective is especially important in drilling environments where actions have immediate cost and safety consequences.

2.2 Digital Transformation and Data-Driven Workflows in Drilling

The drilling industry's digital transformation refers to the incorporation of modern computing and data analytics into drilling operations. Over the past decade, there has been a shift from purely experience-based and physics-based decision-making toward workflows that heavily utilize data-driven models. This transformation has been facilitated by increased sensor instrumentation on rigs, better data infrastructure, and corporate initiatives to leverage big data and AI for efficiency gains. A comprehensive review by Du et al. (2024) chronicles this evolution, noting that multivariate statistical analysis and machine learning are now central to intelligent drilling process monitoring. The traditional siloed approach, where geologists, drilling engineers, and mud loggers previously worked with separate datasets, is giving way to integrated platforms that aggregate data and apply algorithms to extract actionable insights in real time.

A data-driven drilling workflow typically starts with data aggregation from various sources: surface sensors, downhole tools, historical offset well data, and even external data like seismic or geological models. Companies have invested in centralized databases and streaming data platforms to handle the volume and velocity of drilling data. An important aspect of this is ensuring data quality and consistency across sources. Methods for automated data cleaning and normalization have been developed. For example, Abbaszadeh Shahri et al. (2024) introduced an automated approach to normalize large-scale MWD datasets, which is key for creating unified databases that feed machine learning models. By removing tool-specific biases and depth misalignments, such pre-processing enables algorithms to recognize patterns that generalize across different wells and fields.

Once data is aggregated, real-time analytics can be performed. Real-time dashboards at networked operations centers display drilling parameters with algorithm-driven annotations, for instance highlighting abnormal trends or suggesting optimal parameter windows. These systems often incorporate advanced event detection algorithms. As an example, some operators use Bayesian networks or pattern-recognition techniques to detect early signs of kick or lost circulation events from subtle changes in multiple sensor readings. Data-driven Bayesian network approaches have been developed for early kick detection using downhole parameters and historical data, demonstrating improved detection timing and robustness compared with conventional surface-parameter monitoring methods (Dinh 2020). This marks a departure from legacy practices based on rule-of-thumb decisions or single-variable trend monitoring; instead, multiple streams of data are analyzed in concert.

One notable success story in data-driven workflows is the reduction of non-productive time through advanced monitoring. Wang et al. (2023) reported a novel approach to real-time drilling data analysis that significantly reduced NPT in field operations. In their approach, machine learning models were trained to recognize precursor patterns of common drilling problems (such as stuck pipe tendencies or bit wear) and alert the drilling crew with recommended corrective actions. This kind of workflow exemplifies how digital transformation marries domain expertise with data analytics: the models encapsulate knowledge from many wells, and the drilling crew's response strategies are enhanced by this broader insight.

Another pillar of digital transformation is automation and closed-loop control. In fully digital workflows, certain decisions can be automated. For instance, automated geosteering systems ingest LWD log data and adjust the drilling trajectory by sending commands to steering tools, all guided by an AI that aims to keep the well in the target zone. While human oversight remains, the speed and complexity of adjustments achievable by such systems surpass human capabilities. The state of the art includes pilot projects where entire drilling sequences (such as drilling a stand of pipe) are performed automatically by robotic control systems, with AI monitoring downhole conditions to adjust parameters on the fly. These are early steps toward the vision of an autonomous drilling rig.

It is important to note that the digital transformation is not uniform across the industry. Some operators with high-tech fleets and significant R&D investments lead the way, deploying custom AI solutions and advanced sensors, whereas smaller companies may still rely on conventional practices. However, industry-wide initiatives, including collaborations with tech companies and service providers, are helping democratize these technologies. Modern drilling software platforms often come with built-in machine learning capabilities or at least the ability to stream data to cloud AI services for analysis. The trend is clear: data-driven decision support is becoming a standard component of drilling operations, analogous to how measurement-while-drilling became standard for directional drilling over the past few decades.

In summary, the state of the art in digital transformation is characterized by integrated data environments, real-time analytics, predictive modeling of drilling performance, and initial steps towards automation. The industry has accumulated numerous case studies demonstrating value, ranging from improved drilling speed and reduced downtime to better safety. These successes in turn accelerate adoption. As Du et al. (2024) highlight, though challenges remain (data quality, model generalization, and human acceptance), the trajectory is firmly towards more digital, AI-assisted drilling workflows.

2.3 Real-Time Drilling Data Acquisition and Feature Engineering

Effective use of AI in drilling relies on capturing the right data and representing it in ways that machine learning models can interpret, a process known as feature engineering. In real-time drilling operations, data acquisition starts at the rig with a network of sensors measuring diverse aspects of the drilling process. Key measurements include weight on bit (WOB), rotary speed (RPM), torque, pump pressure, flow rate, hook load, standpipe pressure, downhole weight and torque (if available via downhole tools), drilling fluid inlet and return flow, mud logging data (gas readings, cuttings analysis), and LWD logs such as gamma ray or resistivity. These raw measurements form the basis of real-time monitoring.

However, raw data often needs to be transformed into more informative features that highlight the underlying drilling state.

One common engineered feature is mechanical specific energy (MSE), which combines WOB, torque, RPM, and ROP to estimate the energy required to destroy a unit volume of rock. MSE is widely used as a real-time indicator of drilling efficiency; a rising MSE at constant parameters can indicate dull bits or inefficient drilling. Islam et al. (2018) demonstrated the application of MSE and at-bit measurements for geothermal drilling, showing how these engineered metrics correlated with hard rock intervals and could guide adjustments in drilling parameters. The study found that when MSE sharply increased, it often presaged slower drilling or bit wear, validating the use of this feature as a proactive metric in data-driven models.

Feature engineering also involves dealing with the quirks of drilling data. Drilling sensor data is notorious for noise and frequent discontinuities (for example, when pumps are off while making a connection, sensors read zero or drop out). To handle this, real-time systems often apply filters or calculate derivatives. For instance, the rate of change of hook load (dF/dt) can help detect stick–slip or bit bounce events, while the standard deviation of standpipe pressure over a window can signal hole cleaning issues (large fluctuations might mean gas or cuttings influx). The choice of window size for these computations is important: too short and the feature is noisy, too long and it may lag behind the event. Engineers use domain knowledge to set these, or in some advanced systems, adaptive algorithms adjust the window based on the drilling phase (e.g., shorter windows while drilling, longer while tripping).

Another aspect of feature engineering is aligning data on a common reference such as drilled depth or time. Surface measurements are time-indexed, while downhole logs are depth-indexed, which complicates direct integration. Data alignment methods, such as stretch/compress algorithms or sensor fusion via depth tracking, are employed to synchronize events. For example, when a driller pauses pumping (a time-based event) and a pressure drop is seen, aligning that with depth-based data could help identify if the pause corresponded to a known marker (like casing shoe depth). Some AI models sidestep alignment by treating time-series and depth-series separately, but when context from both is needed, careful merging is required. Abbaszadeh Shahri et al. (2024) tackled part of this problem by unifying MWD data from various runs into a normalized depth-scale database. Their automated normalization adjusted for inconsistent depth measurements and sensor delays, which is essential for feeding machine learning models that assume a consistent input structure.

In recent years, automated feature learning through deep learning has gained attention. Instead of manually specifying features, deep neural networks (especially convolutional or recurrent networks) can ingest relatively raw data (with minimal filtering and normalization) and internally learn representations that are useful for prediction. This approach has been successful in many fields and is being applied to drilling. For instance, Altindal et al. (2024) developed an anomaly detection approach on multivariate drilling time series and found that a data-driven feature extraction (via an autoencoder network) could outperform manually selected features for detecting abnormal patterns. However, even with such end-to-end learning, some feature engineering remains at the data preparation stage, for example, scaling of inputs, handling categorical inputs (like bit type or mud type) by encoding them numerically, etc.

The state of the art in feature engineering also includes domain-inspired transformations. For vibration analysis, converting a time-domain signal into a frequency spectrum (via Fourier or wavelet transforms) can highlight specific vibration modes. Alsaihati and Alotaibi (2022), in their work on classifying downhole vibration severity, employed features such as dominant frequency components of torque and acceleration signals to train their machine learning models. This improved the model's ability to distinguish normal drilling from harmful vibration states. Similarly, in bit wear prediction, cumulative drilling distance or number of bit revolutions can be an informative feature derived from time-depth data.

A challenging aspect of real-time drilling data is missing or delayed data. Telemetry dropouts might result in segments of missing downhole measurements. Traditional approaches simply hold the last value or use linear interpolation until data resumes, but this can mislead a reactive algorithm. Recent research, like the TIformer by Ding et al. (2025), provides sophisticated solutions by treating missing data imputation as part of the modeling process. Their transformer-based framework can use known patterns in available data to infer likely values for missing entries, essentially performing feature engineering (filling gaps) on the fly. While such advanced methods are still mostly in research or pilot phases, they represent a trend for resilient feature engineering that can handle real-world data imperfections.

In practice, drilling analytics teams often use a combination of software: high-frequency data is processed by programs running at the rig site for immediate features (such as alarms for pressure spikes), and more complex feature generation may happen on remote servers where computational resources allow for deeper analysis.

2.4 Machine Learning for ROP Prediction and Optimization

Predicting the rate of penetration (ROP), the speed at which the drill bit advances, has long been a focal point for drilling optimization. A higher ROP generally means a well reaches target depth faster, saving time and cost, provided it does not compromise safety or wellbore quality. Traditional ROP models rely on empirical equations calibrated to offset well data. These models can be effective in capturing general trends but often struggle to adapt to changing conditions or to fully exploit the multitude of variables now measured during drilling. Machine learning approaches have therefore been extensively explored to improve ROP prediction accuracy and provide real-time optimization guidance.

Early applications of machine learning for ROP date back at least two decades. Gidh et al. (2012) presented an artificial neural network (ANN) system for drilling parameter optimization that could predict ROP and bit wear trends. Their system took in parameters such as WOB, RPM, bit type, and formation characteristics, and was trained on historical drilling runs to output ROP and an indicator of bit wear. Even with the relatively limited computational power of that time, the ANN showed an ability to generalize and provided better ROP estimates than linear models, especially in sections where lithology changed rapidly. This work demonstrated the promise of machine learning, but also highlighted challenges: the "black box" nature of neural networks made it hard for drilling engineers to trust the predictions without understanding the reasoning, and the models required substantial data for training.

In recent years, the literature shows a proliferation of techniques for ROP modeling: from classical machine learning algorithms like support vector machines and random forests to modern deep learning architectures. Many studies have validated that given enough quality data, these models can achieve high predictive accuracy. For example, Aoun and Madarász (2022) developed an ensemble of machine learning models (including gradient boosting and deep neural nets) to predict ROP in a geothermal well and reported a notable match with actual drilling data in the FORGE project. Because geothermal wells often drill through heterogeneous rock, they included indicators of rock type and temperature in the feature set, allowing the model to adjust ROP predictions when transitioning from, say, sedimentary to igneous formations. This flexibility is an advantage of machine learning models as they can incorporate a wide range of input features without needing an explicit physical formula relating them to ROP.

Machine learning models for ROP are also being used in optimization mode. Instead of just forecasting what ROP will be, the goal is to recommend the set of controllable parameters (WOB, RPM, flow rate, etc.) that would maximize ROP under current conditions. Approaches to this include training a model on past data to output ROP, and then using that model as a surrogate in an optimization algorithm to find the best parameters. Hegde et al. (2018) exemplified this hybrid approach: they combined a deterministic drilling model with a machine learning model that learned the residual differences. This hybrid ROP model could honor known physical effects while the machine learning component adjusted for unmodeled factors. Using this model, they performed what-if simulations to identify optimal parameters for maximizing ROP. Notably, their results showed improved ROP predictions and provided guidance that in field trials led to ROP improvements of 10–20% in certain intervals by adjusting practices such as weight on bit and flow rates.

One practical concern in ROP modeling is generalization as models sometimes perform well on the wells they were trained on but falter in new fields with different geology. To mitigate this, transfer learning and continual learning strategies are being used. For instance, a model might be pre-trained on a large dataset from many wells to capture general drilling response trends, and then fine-tuned on a smaller dataset from the specific field of interest. The work by Atashnezhad et al. (2021) on a drilling optimization system in geothermal wells essentially followed this approach: they built a broad model using various geothermal wells' data, then adapted it to the target well. Once calibrated, their system provided real-time recommendations for adjusting drilling parameters to improve ROP, and field implementation reportedly yielded better ROP than neighboring wells drilled without the system.

Modern machine learning models have shown high accuracy in ROP predictions when sufficient data is available, but explainability remains a concern. Recent research has started to incorporate model interpretability techniques, such as SHAP values or decision tree surrogates, to help engineers understand which inputs are most influencing the ROP predictions. This builds trust and aids in knowledge extraction, for instance, confirming that the model correctly recognizes known effects and discovering new patterns.

2.5 Drilling Dynamics and Vibration Control

Drilling dynamics encompasses the various modes of movement and forces in the drill string and bottom-hole assembly (BHA), including rotation, axial oscillation, and lateral vibrations. Uncontrolled vibrations, such as stick–slip, bit bounce, and whirl, can damage tools, reduce ROP, and even cause wellbore instability. Controlling these dynamics is essential for safe and efficient drilling. Traditionally, drillers mitigate vibrations by adjusting surface parameters or using downhole tools. However, recognizing and quantifying downhole vibrations in real time is challenging because direct measurements are limited and surface indications can be subtle. AI techniques have emerged as powerful aids for detecting, classifying, and even predicting harmful vibration states so that timely control actions can be taken.

One state-of-the-art application is the use of pattern recognition and classification algorithms on surface sensor data to infer downhole vibration severity. Alsaihati and Alotaibi (2022) developed models using three different machine learning techniques, including decision trees and support vector machines, to determine the severity of lateral and torsional vibrations while drilling surface-hole sections. They trained these models on data labeled with vibration severity levels derived from high-speed downhole sensor recordings and post-run analysis. Input features came from surface measurements such as fluctuations in drill string torque and hook load variability, as well as any downhole shock sensor readings available. The machine learning models achieved high accuracy in classifying vibration severity and could do so continuously while drilling. This capability allows a driller or an automated system to dynamically adjust drilling parameters. For example, if the model flags an increasing torsional oscillation trend, the driller might adjust the top drive RPM or WOB to dampen it, preventing the severe stick–slip that can cause tool failure.

Beyond classification, machine learning is used in anomaly detection for drilling dynamics. Not all vibrations manifest in easily recognized patterns. Some events, like an emerging bottom-hole assembly whirl, might not have a clear signature in standard measurements until it becomes severe. Unsupervised learning methods that learn the normal pattern of drilling dynamics and then detect deviations have been employed. Altindal et al. (2024) proposed an anomaly detection framework on multivariate drilling data that can capture unusual dynamic behaviors without explicit labels. By training an autoencoder on normal drilling states, their system could detect the onset of abnormal vibration modes as anomalies when the reconstruction error spiked. This approach is valuable for catching the early signs of a problem that might not yet be classified as severe but is trending in that direction.

Rykov (2024) provides an overview of how mechanical vibrations contribute to BHA component failures, highlighting the importance of continuous monitoring. In terms of AI, machine learning models can incorporate domain insights, for instance, knowing that severe lateral vibrations correlate with specific accelerometer signatures and tool failures.

In summary, the state of the art in drilling dynamics and vibration control uses AI for detection and classification of harmful vibrations, unsupervised anomaly detection to catch emerging issues before they escalate, short-term prediction of dynamic dysfunctions to enable proactive control, and intelligent advisory systems that integrate machine learning outputs with drilling engineering knowledge to recommend mitigations. These AI-driven capabilities help extend tool life, maintain higher drilling performance, and reduce the risk of severe vibration-related incidents.

2.6 Predictive Maintenance and Tool-Health Monitoring

Drilling operations involve complex assemblies of mechanical and electronic tools that operate under harsh conditions. Unexpected failure of these tools can lead to costly non-productive time and even jeopardize the well. Predictive maintenance aims to anticipate tool failures before they happen, so that maintenance can be performed at a convenient time rather than after a catastrophic failure. AI techniques, especially those for temporal pattern recognition and anomaly detection, have become central to building effective predictive maintenance systems in drilling.

Another promising trend is the use of digital twins for predictive maintenance. A digital twin is a virtual model of a physical tool that runs in parallel with the operation, updated in real time with data. AI is used within the twin to predict failure modes. Mikołajewska et al. (2025) discussed how generative AI can enhance digital twins for fault diagnosis. For drilling, a digital twin of a mud motor would simulate its wear and tear based on drilling parameters and compare its simulated sensor outputs with actual measurements. Any divergence could indicate an emerging issue. The generative aspect might involve simulating realistic failure progression scenarios. By integrating physics-based models of motor wear with AI that calibrates those models to actual usage, the twin provides a more interpretable insight.

Machine learning has also been applied to specific maintenance issues like drill bit wear estimation. Traditionally, bit wear is assessed after pulling out. AI models have been trained to infer bit wear state from drilling data trends. Features such as ROP reduction over footage, increased torque, and MSE have been input into regression models to estimate the bit dull condition in real time. These models can help decide whether continuing to drill with the current bit is economical or if a trip should be made to change it. Motahhari et al. (2010) introduced an approach to improve drilling efficiency by integrating bit parameters and performance data, which can be seen as a precursor to current machine learning-based bit wear predictors.

It is also instructive to consider how similar predictive maintenance systems have succeeded in other industries, as drilling is now emulating those successes. For example, deep learning models have been used in aerospace to predict remaining useful life of jet engines, and Biggio et al. (2023) used deep learning for battery discharge prediction in electric vehicles. Techniques from those domains are being transferred to drilling tools. The battery discharge prediction is analogous to tracking sensor drift or battery life of downhole electronics, by monitoring how a tool's performance indicators change over time, an AI model can forecast when it will no longer perform to specification.

References

Abbaszadeh Shahri, A., Shan, C., Larsson, S., & Johansson, F. (2024). Normalizing large scale sensor based MWD data: An automated method toward a unified database. Sensors, 24(4), 1209.

Alsaihati, A., & Alotaibi, B. (2022). Determining severity of lateral and torsional downhole vibrations while drilling surface holes using three machine learning techniques. SPE Journal, 27(3), 1493–1503.

Altindal, M. C., Nivlet, P., Tabib, M., Rasheed, A., Kristiansen, T. G., & Khosravanian, R. (2024). Anomaly detection in multivariate time series of drilling data. Geoenergy Science and Engineering, 237, 212778.

Aoun, M. A. B., & Madarász, T. (2022). Applying machine learning to predict the rate of penetration for geothermal drilling located in the Utah FORGE site. Energies, 15(12), 4288.

Atashnezhad, A., Akhtarmanesh, S., Hareland, G., & Al Dushaishi, M. (2021). Developing a drilling optimization system for improved overall rate of penetration in geothermal wells. In ARMA US Rock Mechanics/Geomechanics Symposium. ARMA.

Biggio, L., Bendinelli, T., Kulkarni, C., & Fink, O. (2023). Ageing-aware battery discharge prediction with deep learning. Applied Energy, 346, 121229.

Ding, Z., Chen, Y., Wang, H., Wang, X., Zhang, W., & Zhang, Y. (2025). TIformer: A transformer-based framework for time-series forecasting with missing data. In Australasian Database Conference (pp. 71–84). Springer.

Dinh, N. M. (2020). Early kick detection using data-driven Bayesian network: Model development and experimental testing (Doctoral dissertation, Memorial University of Newfoundland).

Du, S., Huang, C., Ma, X., & Fan, H. (2024). A review of data-driven intelligent monitoring for geological drilling processes. Processes, 12(11), 2478.

Gidh, Y., Purwanto, A., & Smith Bits. (2012). Artificial neural network drilling parameter optimization system improves ROP by predicting and managing bit wear. In SPE Intelligent Energy International Conference and Exhibition. SPE.

Hegde, C., Soares, C., & Gray, K. (2018). Rate of penetration modeling using hybrid models: Deterministic and machine learning. In Unconventional Resources Technology Conference (pp. 3220–3238).

Islam, N., Vijapurapu, R., Jones, M., McLennan, J., Moore, J., Rickard, W., Balamir, O., & Vagnetti, R. (2018). Application of mechanical specific energy and at-the-bit measurements for geothermal drilling applications. In ARMA US Rock Mechanics/Geomechanics Symposium. ARMA.

Mikołajewska, E., Mikołajewski, D., Mikołajczyk, T., & Paczkowski, T. (2025). Generative AI in AI-based digital twins for fault diagnosis for predictive maintenance in Industry 4.0/5.0. Applied Sciences, 15(6), 3166.

Motahhari, H. R., Hareland, G., & James, J. A. (2010). Improved drilling efficiency technique using integrated PDM and PDC bit parameters. Journal of Canadian Petroleum Technology, 49(10), 45–52.

Rykov, Y. (2024). Drilling and logging equipment reliability in a downhole vibration environment. Journal of Petroleum Technology.

Wang, J., Sajeev, S. K., Ozbayoglu, E., Baldino, S., Liu, Y., & Jing, H. (2023). Reducing NPT using a novel approach to real-time drilling data analysis. In SPE Annual Technical Conference and Exhibition. SPE.

3 Geothermal Drilling Rate of Penetration Forecasting

3.1 Background and Related Work

Drilling performance optimization remains one of the most critical factors influencing the success and cost-effectiveness of geothermal and CCS operations. Achieving consistent performance under highly variable subsurface conditions requires accurate, reliable, and adaptive predictive models. Historically, drilling optimization relied heavily on empirical models, statistical regressions, and operator expertise. While these approaches provided insights into key performance metrics, such as ROP and shock dynamics, they often fell short in capturing the nonlinear, multi-dimensional relationships in drilling data. Furthermore, static models based on idealized assumptions struggle to adapt to real-time operational changes, sensor noise, and equipment degradation.

3.1.1 Drilling Performance Optimization in Geothermal Wells

Drilling operations in geothermal wells represent one of the most technically challenging and cost-intensive components of sustainable energy extraction. The inherently harsh drilling environments, characterized by elevated temperatures, abrasive rock formations, and high-pressure subsurface conditions, pose barriers to efficiency and cost-effectiveness. Optimization of drilling performance in geothermal wells is critical not only for reducing operational expenditures but also for minimizing NPT, equipment wear, and environmental impacts. At the core of drilling performance lies ROP, which serves as a critical indicator of drilling efficiency. Enhancing this metric has been the focus of research efforts, with methodologies ranging from empirical optimization techniques to advanced ML-based predictive models.

Traditional drilling optimization techniques have relied heavily on empirical models and heuristic approaches. Metrics such as MSE have been used to evaluate energy efficiency during drilling operations by quantifying the energy required to drill through a unit volume of rock. Rickard et al. (2019) demonstrated the utility of MSE in identifying inefficiencies caused by mechanical dysfunctions such as stick–slip vibrations, bit whirl, and excessive bit wear. These inefficiencies often result in suboptimal drilling conditions where increases in WOB or rotational speed fail to yield proportional improvements in ROP. While MSE models offer valuable diagnostic capabilities, they are inherently limited by their reliance on static assumptions and simplified mathematical relationships that often fail to account for the dynamic interactions among drilling parameters.

The optimization of drilling parameters, including WOB, RPM, torque, and mud flow rate, has traditionally been approached through

C. Urdaneta et al., *Deep Learning Applications for Drilling Performance: Forecasting, Diagnostics, Telemetry, and Predictive Maintenance*, Advances in Oil and Gas Exploration & Production,
https://doi.org/10.1007/978-3-032-26648-4_3

manual tuning and operator expertise. Field engineers frequently rely on experience-based adjustments to fine-tune these parameters in real time. However, such manual approaches are reactive, inconsistent, and susceptible to human error, particularly in geothermal wells where drilling conditions can change rapidly. For instance, studies by Islam et al. (2018) have shown that while increasing WOB can enhance ROP in certain formations, excessive weight often results in bit wear and diminishing returns, a phenomenon known as bit founder. This nonlinearity shows the limitations of manual parameter optimization and the need for more dynamic, data-driven approaches.

In response to these limitations, recent advancements in automation and real-time data acquisition have paved the way for more sophisticated optimization techniques. MWD tools and surface sensors now provide continuous streams of high-resolution data on parameters such as downhole pressure, vibration, torque, and WOB. These data streams enable the application of real-time optimization frameworks that can dynamically adjust drilling parameters based on current operating conditions. The integration of these tools into predictive models has been shown to reduce NPT and improve drilling efficiency. For example, Atashnezhad et al. (2021) developed an optimization algorithm that accounts for unconfined compressive strength and temperature limitations to maximize ROP while preventing excessive bit wear.

Despite these advancements, geothermal wells present unique challenges that complicate the application of traditional optimization techniques. High-temperature environments can accelerate bit wear, resulting in nonlinear performance degradation over time. These factors make it difficult to rely on static optimization models, as the drilling environment continuously evolves with depth and time. Furthermore, the high abrasiveness of geothermal formations increases the likelihood of mechanical dysfunctions, which can disrupt the expected relationships between drilling parameters and ROP.

The rise of AI and ML has introduced transformative possibilities for drilling performance optimization. Unlike traditional empirical models, ML algorithms can identify complex, nonlinear relationships among drilling parameters without requiring explicit mathematical formulations. Techniques such as artificial neural networks (ANN), random forests, and gradient boosting have been successfully applied to predict ROP and optimize drilling parameters in geothermal wells. For example, Aoun and Madarász (2022) utilized ML algorithms to predict ROP in geothermal drilling using data from the Utah FORGE site. Their study demonstrated improvements in prediction accuracy compared to traditional regression-based models, highlighting the potential of ML for geothermal drilling optimization.

Transformer-based deep learning (DL) models represent the latest frontier in this domain. These models, initially developed for natural language processing tasks, excel at capturing long-term dependencies and sequential patterns in time-series data. In the context of geothermal drilling, transformers offer unique advantages in analyzing large volumes of real-time sensor data to predict ROP with high temporal granularity. Studies by Phelan et al. (2022) demonstrated that ML models can effectively predict formation properties from drilling data, accounting for geological variability and sensor noise. While their work did not directly apply to ROP forecasting with transformers, it shows how ML can uncover subtle patterns in drilling datasets, which is essential for anticipating performance issues and informing proactive adjustments to drilling parameters.

However, the application of transformer models in geothermal drilling optimization is not without challenges. The quality and consistency of drilling datasets remain barriers, as sensor noise, irregular sampling intervals, and incomplete measurements can compromise model performance. Additionally, the deployment of predictive models in live drilling workflows introduces technical constraints such as computational latency, synchronization with real-time sensor streams, and the need for a confidence score.

Research efforts such as those by Motahhari et al. (2010) and Gidh et al. (2012) have highlighted the importance of integrating predictive models with real-time control systems to achieve closed-loop optimization. In such systems, model predictions are continuously fed into drilling automation frameworks, enabling adaptive adjustments to parameters such as WOB and RPM. This integration not only enhances drilling efficiency but also minimizes operator intervention, reducing the potential for human error.

3.1.2 ML Applications in Drilling Optimization

The application of ML techniques in drilling optimization has gained traction due to the complexity and dynamic nature of drilling operations. Traditional empirical models and rule-based optimization frameworks often fall short when capturing the nonlinear relationships between drilling parameters, geological variability, and operational outcomes. ML models, particularly those capable of handling large datasets with high temporal resolution, offer enhanced predictive capabilities that complement traditional engineering approaches.

In recent years, supervised learning models such as ANNs, support vector machines, and random forests have been widely applied in drilling optimization tasks, including predicting drilling parameters, identifying dysfunctions, and optimizing real-time control systems. For example, Gidh et al. (2012) developed an ANN-based system for drilling parameter optimization, focusing on reducing operational costs and extending bit life. This system leveraged historical drilling data to predict optimal WOB and RPM settings, demonstrating improved performance and cost efficiency in field trials.

Another key advancement comes from the use of MSE as a proxy for drilling efficiency, which has been effectively integrated with ML algorithms. Islam et al. (2018) employed MSE combined with downhole measurements as real-time tools for drilling optimization. By analyzing the relationship between drilling parameters and MSE, their approach facilitated adaptive parameter adjustments to improve drilling performance. Similarly, Atashnezhad et al. (2021) proposed an optimization algorithm that combined ML models with rock compressive strength and bit wear metrics to maximize penetration rates while maintaining bit integrity.

Beyond conventional supervised learning techniques, hybrid models that combine deterministic drilling equations with data-driven approaches have shown promise. Hegde et al. (2018) demonstrated the potential of hybrid models in ROP prediction, where deterministic models provided a baseline prediction refined by ML algorithms. This approach reduced prediction errors and improved interpretability, a critical factor in operational environments where transparency is essential for decision-making.

DL, a subset of ML, has further improved drilling optimization by enabling models to learn hierarchical representations of drilling data. Recurrent neural networks (RNNs), specifically long short-term memory (LSTM) networks, have been widely applied for time-series forecasting of drilling parameters. However, RNNs suffer from limitations such as vanishing gradient issues and computational inefficiency when processing long time-series data. Transformers, introduced by Vaswani et al. (2017), have emerged as an alternative to traditional recurrent and convolutional architectures for modeling long-range temporal dependencies in drilling time-series data.

Transformers are suitable for analyzing drilling data streams from MWD tools, where parameters such as WOB, RPM, torque, and pressure exhibit complex temporal relationships. By dynamically adjusting their attention to different time intervals and parameter combinations, transformers can extract meaningful patterns from noisy, high-dimensional data streams. This ability is particularly useful for addressing drilling dysfunctions such as stick–slip vibrations, bit whirl, and shock dynamics, which are challenging to predict using traditional models.

3.1.3 Advances in Transformer Models for Time Series Forecasting

Transformer models have emerged as a novel architecture for time series forecasting, offering improvements over traditional RNNs and CNNs. Initially introduced by Vaswani et al. (2017) for natural language processing tasks, transformers improved sequential data analysis by replacing recurrence with self-attention mechanisms. These mechanisms enable transformers to efficiently capture both short-term dependencies and long-range relationships within sequential data, addressing the limitations of RNNs and their variants, such as LSTMs and gated recurrent units (GRUs).

Time series forecasting, especially in industrial domains such as drilling operations, presents unique challenges. Data collected from MWD tools are inherently noisy, sparse, and irregularly sampled. Moreover, drilling parameters such as WOB, RPM, torque, and MSE interact in complex, nonlinear ways. Traditional time series models, including autoregressive integrated moving average (ARIMA) and LSTMs, struggle to capture these nonlinear dependencies over varying temporal scales. Transformers, however, excel in such environments due to their self-attention mechanism, which allows them to weigh the importance of different time steps in the data sequence dynamically.

One of the key innovations in transformers for time series forecasting is the self-attention mechanism. Unlike RNNs, where dependencies between time steps are modeled sequentially, transformers process time steps simultaneously. Each time step attends to all other steps, enabling the model to learn relationships across the entire dataset without the risk of vanishing gradients or long-term dependency degradation. This parallel processing enhances computational efficiency, allowing transformers to handle longer time series sequences with greater accuracy and scalability.

Another advancement is the positional encoding technique, which enables transformers to retain the sequential order of time series data. Since transformers lack a sense of order due to their parallel structure, positional encodings are added to the input embeddings to provide information about the relative or absolute positions of time steps. This feature ensures that temporal dependencies are preserved during model training and inference.

Recent transformer variants, such as Informer (Zhou et al. 2021) and Autoformer (Wu et al. 2021), have been tailored for time series forecasting. Informer introduces a sparse self-attention mechanism that reduces computational overhead, making it more suitable for long time series data. This optimization allows Informer to efficiently process sequences by focusing computational resources on the most relevant time points. Autoformer, on the other hand, incorporates series decomposition techniques and an auto-correlation mechanism to better capture long-term temporal dependencies and periodic patterns. These adaptations make Autoformer effective in modeling the seasonal and trend components often present in time series data. These developments show the utility of transformer architectures in addressing the challenges posed by sequential data.

The Dynaformer model, developed by Biggio et al. (2023), is an example of leveraging transformers for industrial time series data. Initially designed for battery discharge prediction, Dynaformer segments time series data into smaller subsequences, treating each subsequence as a token in the transformer architecture. This segmentation reduces computational complexity and enhances the model's ability to focus on relevant time intervals while disregarding irrelevant noise. A similar approach can be adapted to drilling datasets, where the segmentation of drilling parameters into manageable tokens can optimize predictive performance.

Furthermore, the integration of multi-head self-attention in transformer models enhances their ability to simultaneously focus on different aspects of drilling data, enabling a more comprehensive analysis of complex interactions among operational parameters. Each attention head operates independently, capturing unique relationships within the data. For instance, one attention head might prioritize identifying patterns in WOB variations, while another could focus on the influence of RPM on the ROP. This parallel attention mechanism ensures that diverse temporal and parameter-specific dependencies are learned and represented within the model. As a result, transformer models are better equipped to capture the intricate, nonlinear relationships that govern drilling performance, leading to improved predictive accuracy.

While transformer models have demonstrated superior performance in time series forecasting, their adoption in real-time drilling workflows presents several challenges. Computational latency remains a concern, especially when predictions must be delivered with minimal delays. Strategies such as model optimization, efficient deployment architectures, and quantization techniques have been proposed to address these constraints. For example Girase et al. (2023) introduced RAFTformer, a real-time action forecasting transformer designed to reduce inference time without compromising predictive accuracy. Their approach shows the importance of low-latency architectures in enabling real-time decision-making workflows.

In addition to computational concerns, drilling datasets often exhibit irregular sampling intervals, sensor noise, and missing values, which complicate the training and deployment of predictive models. Ding et al. (2025) proposed the Time-series Imputation Transformer (TIformer), which integrates imputation techniques directly within the transformer framework. Their model effectively handles missing data in time series datasets, improving reliability and generalization across diverse operational conditions.

Another area of advancement lies in transfer learning with transformers for time series data. Transfer learning enables pre-trained transformer models to be fine-tuned on specific datasets, reducing training time and improving predictive accuracy. In drilling applications, this approach allows models trained on historical wells to be rapidly adapted to new wells, leveraging prior knowledge while accounting for site-specific geological and operational conditions.

3.1.4 Challenges in ROP Forecasting

Accurate prediction of ROP remains a central challenge in drilling optimization due to the complex interplay of geological, mechanical, and operational factors. Each of these aspects introduces uncertainties and nonlinearities, complicating the development of reliable predictive models. While advancements in data analytics and ML have provided promising tools for ROP prediction, these methods still face several limitations when confronted with the dynamic nature of real-world drilling environments.

Geological factors affect ROP due to the heterogeneity and unpredictability of subsurface formations. Rock strength, abrasiveness, porosity, permeability, and compressibility vary considerably, both vertically and laterally, across drilling trajectories. Hard, abrasive formations often reduce ROP and cause accelerated bit wear, while softer, porous formations generally allow for higher penetration rates. Additionally, abrupt changes in lithology and the presence of interbedded formations introduce variability in drilling performance, disrupting the consistency of predictive models. Geological uncertainties are compounded by the limitations of pre-drill geological surveys, which may lack the resolution or accuracy needed to inform real-time drilling optimization effectively. Rickard et al. (2019) showed these challenges in the Utah FORGE dataset, where formation heterogeneity contributed to variations in drilling efficiency across different well sections.

Mechanical factors also play a role in ROP prediction, with WOB, RPM, torque, and drilling fluid properties being key contributors. The MSE metric, as discussed by Rickard et al. (2019), provides a framework for assessing energy efficiency during drilling operations. However, this metric does not fully account for unmeasured parameters such as bit wear, downhole vibrations, and pressure gradients ahead of the drill bit. Mechanical dysfunctions, including stick–slip, bit balling, whirl vibrations, and cutter wear, can further degrade drilling efficiency. Each of these dysfunctions introduces nonlinear responses between WOB, RPM, and ROP, requiring real-time adjustments to drilling parameters. Studies by Islam et al. (2018) and Atashnezhad et al. (2021) emphasize the need for dynamic optimization strategies to counteract mechanical dysfunctions and maintain optimal ROP throughout the drilling process.

Operational factors encompass drilling practices, rig capabilities, and real-time decision-making by drillers and engineers. Operational inefficiencies often arise from suboptimal parameter settings, delayed adjustments, and reliance on static drilling models. Additionally, limitations in sensor accuracy, data transmission delays, and poor integration of real-time data streams pose hurdles to effective ROP prediction. MWD tools provide continuous feedback on key drilling parameters, but sensor noise, intermittent malfunctions, and irregular sampling intervals often degrade data quality (Abbaszadeh Shahri et al. 2024). Furthermore, operational workflows may involve conservative parameter settings to minimize risks, even at the expense of reduced ROP. This trade-off between operational safety and drilling efficiency remains a challenge in many drilling projects.

Another operational factor influencing ROP prediction is data sparsity and imbalance in drilling datasets. While modern drilling operations generate large volumes of time-series data, the distribution of these datasets is often uneven across different operational conditions. Specific drilling dysfunctions or extreme geological transitions may be underrepresented in the data, leading to biases in predictive models. Data preprocessing and augmentation techniques, such as time warping and noise injection, have been explored in the broader time-series literature to address similar issues of data imbalance and sparsity (Iglesias et al. 2023). However, their application in drilling datasets remains limited, and the effectiveness of these techniques in real-world scenarios is still under investigation.

3.2 Dataset Description and Preparation

3.2.1 Utah FORGE Dataset Overview

The Utah FORGE represents a key initiative aimed at advancing enhanced geothermal systems technologies through large-scale experimental drilling projects. Supported by the U.S. Department of Energy, it serves as a controlled field laboratory for investigating drilling techniques, reservoir stimulation, and subsurface resource extraction under real-world geothermal conditions. Situated in Milford, Utah, the site features a granitic basement rock characterized by high temperatures, thermal gradients, and low natural permeability, making it an ideal environment for testing innovative drilling methodologies and predictive optimization frameworks.

The dataset derived from this project is one of the most comprehensive publicly available geothermal drilling datasets and serves as the foundation for the current research. It encompasses time-series data collected from multiple wells, including a pilot well, seismic monitoring wells, and a highly deviated injection well (Fig. 3.1). These wells represent diverse operational and geological conditions, offering a robust basis for evaluating data-driven predictive models. Among them, the pilot well (58-32) is of particular interest due to its high-resolution drilling data, which captures the full lifecycle of drilling phases and provides granular insights into performance metrics such as ROP and mechanical dysfunctions.

The dataset consists of continuous measurements recorded through downhole tools and surface sensors during drilling operations. These measurements include parameters such as WOB, RPM, torque, downhole pressure, mud flow rates, and vibration. These variables are known to exhibit complex, nonlinear interdependencies with drilling performance metrics, as demonstrated in earlier studies (Rickard et al. 2019;

Fig. 3.1 Utah FORGE Wells (Rickard et al. 2019)

Islam et al. 2018). Additionally, the dataset captures transient effects caused by changes in operational settings, mechanical wear on drilling equipment, and abrupt geological transitions, all of which affect drilling efficiency.

One of the dataset's strengths lies in its temporal granularity. Time-series measurements are recorded at intervals ranging from one second to several minutes, depending on the drilling phase and data acquisition configuration. However, the varying resolution introduces preprocessing challenges, as inconsistencies in time intervals and sensor noise can degrade model performance if not addressed systematically. Data preprocessing routines typically involve resampling to achieve consistent time intervals, normalizing parameters across different wells, and removing outliers introduced by sensor malfunctions or data logging errors.

Metadata accompanying the dataset further enriches its interpretability, providing contextual information about well design specifications, lithology logs, equipment configurations, and operator notes. These metadata elements are critical for identifying relationships between drilling parameters and subsurface geology, as well as for validating predictive insights generated by ML models. For instance, lithology logs enable an understanding of depth-specific changes in rock compressibility and abrasiveness, which directly affect drilling efficiency and equipment wear (Gwynn et al. 2018; Rickard et al. 2019).

A portion of the Utah FORGE dataset is suitable to exploring the relationship between WOB, RPM, torque, and ROP. As highlighted by Rickard et al. (2019), deviations from expected linear relationships among these parameters often signal the onset of drilling dysfunctions such as stick–slip vibrations, whirl motion, or bit balling. For example, excessive axial force without a corresponding increase in drilling speed can indicate bit wear or mechanical inefficiency, both of which are reflected in elevated MSE values. These insights emphasize the importance of developing predictive models capable of capturing such nonlinearities, especially under the variable conditions encountered in geothermal wells.

The dataset also includes specific case studies and event logs detailing notable occurrences during drilling operations, such as equipment failures, drilling halts, or abrupt changes in operational strategies. These logs provide qualitative data that complement the quantitative time-series measurements, offering a more holistic perspective on drilling performance. Studies by Islam et al. (2018) and Atashnezhad et al. (2021) have highlighted the importance of integrating contextual metadata with numerical drilling parameters to enhance the accuracy and interpretability of predictive models.

Previous studies leveraging the Utah FORGE dataset have demonstrated its utility in exploring predictive modeling approaches for drilling optimization. For example, Rickard et al. (2019) demonstrated the correlation between MSE and drilling efficiency, highlighting the dataset's potential for identifying operational inefficiencies. These findings highlight the dataset's value as a resource for developing and validating advanced AI-driven predictive models.

In the context of this research, the Utah FORGE dataset serves as the foundation for developing transformer-based predictive models aimed at ROP forecasting. The high temporal resolution, diversity of drilling parameters, and availability of contextual metadata provide an ideal environment for training and validating advanced ML architectures. Furthermore, the dataset's coverage of multiple wells enables cross-validation of model performance across varying geological formations and operational conditions.

3.2.2 Key Drilling Parameters in the FORGE Dataset

The FORGE dataset has a wide range of drilling parameters critical for understanding and optimizing drilling performance in geothermal environments. These parameters are derived from high frequency measurements collected by downhole and surface sensors during various drilling phases. They represent the foundation upon which predictive models, such as transformer based architectures, are developed and validated. Understanding the significance of these parameters, their interdependencies, and their variability across geological formations is essential for accurately predicting key performance indicators like the ROP and MSE.

One of the most important parameters in the dataset is WOB, which measures the axial force applied to the drill bit. WOB directly influences the cutting efficiency of the bit, with optimal ranges often varying based on lithology, bit design, and operational objectives. Excessive axial force can lead to bit wear, inefficient energy transfer, and mechanical dysfunctions such as bit whirl or stick slip vibrations. Conversely, insufficient axial force may prevent the bit from effectively breaking the rock, resulting in low penetration rates. Previous research, including studies from the Utah FORGE wells, has shown that deviations from optimal WOB values can disrupt the expected linear relationship with drilling speed, emphasizing the need for continuous parameter adjustments (Rickard et al. 2019; Gwynn et al. 2018).

Another important parameter is RPM, which measures the rotational speed of the bit. Alongside WOB, RPM plays a role in determining drilling efficiency. Studies have shown that the interaction between WOB and RPM is not always linear, particularly in high temperature geothermal formations where bit ball wear and thermal stress affect performance. Excessive rotational speeds can accelerate bit wear and induce harmful vibrational modes, such as torsional stick slip or lateral whirl vibrations. Conversely, insufficient rotational speeds may reduce the energy transferred to the rock face, limiting drilling progress (Islam et al. 2018). These complexities require real time monitoring and predictive control mechanisms to maintain optimal rotational speeds across varying subsurface conditions.

Torque, which measures the rotational force applied to the drill string, serves as a complementary parameter to WOB and RPM. Torque is influenced by downhole friction, rock strength, bit design, and fluid flow dynamics. Excessive torque can indicate increased resistance from the rock formation, suboptimal drilling parameters, or mechanical dysfunctions. Real time analysis of torque trends, particularly in conjunction with WOB and RPM data, can reveal insights into bit rock interaction efficiency and early signs of dysfunctions such as bit balling or differential sticking.

Mud flow rate is another fundamental parameter included in the FORGE dataset. Mud serves multiple roles during drilling, including cuttings removal, cooling the drill bit, maintaining wellbore stability, and preventing formation fluid influx. Flow rate influences hole cleaning efficiency and directly impacts MSE and ROP. Insufficient flow rates can lead to poor cuttings transport, causing bit balling and increased mechanical resistance. On the other hand, excessive flow rates can result in hydraulic pressure imbalances and energy loss through turbulent fluid dynamics. Balancing mud flow rate with WOB, RPM, and torque is essential for maintaining drilling efficiency, particularly in geothermal wells where thermal gradients exacerbate fluid dynamics challenges.

Downhole pressure measurements are also a key part of the dataset, providing insights into wellbore stability and fluid behavior during drilling. Pressure data, particularly in geothermal formations, can indicate zones of high porosity or fluid influx, both of which pose operational risks. Downhole pressure variations also affect the formation's response to mechanical stresses, influencing parameters like MSE and drilling efficiency.

MSE represents a derived parameter that quantifies the energy required to drill through a unit volume of rock. It serves as an indicator of drilling efficiency and is derived from WOB, RPM, torque, and ROP data. High MSE values often signal drilling inefficiencies, such as worn bits, poor weight distribution, or excessive energy losses due to vibrations. Studies utilizing the Utah FORGE dataset have demonstrated the utility of MSE in diagnosing drilling dysfunctions and optimizing energy transfer efficiency (Rickard et al. 2019).

Vibration metrics, including axial, torsional, and lateral vibrations, are also extensively recorded in the dataset. These vibrations can disrupt optimal drilling conditions and cause damage to drilling equipment. Stick slip vibrations have been identified as a common issue in high temperature geothermal wells. Excessive vibrations often manifest as fluctuations in WOB, torque, and RPM measurements. The ability to predict and mitigate vibration related dysfunctions is necessary for ensuring smooth drilling operations and minimizing nonproductive time.

Additional parameters, such as temperature, depth, and drilling speed, further enrich the dataset. Temperature data are particularly valuable in geothermal drilling, where thermal stresses affect both drilling equipment and fluid dynamics. Depth serves as a contextual variable, providing a spatial reference for correlating drilling parameters with geological formations. Drilling speed, expressed as the ROP, serves as the primary performance metric and is influenced by the dynamic interplay of all the parameters.

3.2.3 Data Preprocessing and Feature Engineering

Data preprocessing begins with addressing missing values, a common issue in drilling datasets due to sensor malfunctions or transmission errors. Techniques such as interpolation or imputation are employed to estimate and fill these gaps, ensuring continuity in time-series data. For instance, linear interpolation can estimate missing sensor readings by considering adjacent time points, preserving the temporal dynamics essential for accurate modeling.

Outlier detection and removal constitute another aspect of preprocessing. Drilling operations often encounter anomalies caused by sudden equipment malfunctions or unexpected geological formations. Statistical methods, such as z-score analysis, identify these outliers by measuring deviations from the mean or interquartile range. Once detected, outliers can be addressed through methods like Winsorization or, if appropriate, removal, preventing skewed model training.

Normalization of continuous variables is essential to harmonize the scales of different parameters, such as WOB, RPM, and torque. By scaling these variables to a standard range, typically between zero and one, normalization ensures that no single feature disproportionately influences the model. This step is important for algorithms sensitive to feature magnitude, facilitating efficient convergence during training.

Feature engineering in this study focused on creating new, informative variables that enhance the model's ability to capture complex relationships and dependencies within the drilling dataset. Instead of relying solely on raw sensor measurements, additional features were derived through pairwise linear and polynomial combinations of key parameters, such as WOB, flow rate, and differential pressure. These engineered variables aimed to represent higher-order interactions and provide the model with enriched context for predicting both ROP and downhole shock. Additionally, variable importance analysis was employed to identify which derived features contributed most to the model's predictive accuracy, ensuring that critical relationships were prioritized during training. By transforming raw sensor data into more meaningful representations, this approach improved the model's ability to generalize across varying operational and geological conditions.

Dimensionality reduction techniques, such as Principal Component Analysis, are employed to distill the feature set by identifying and retaining the most variables. This reduction mitigates the risk of overfitting, decreases computational load, and enhances model interpretability by focusing on the features that contribute most to variance in the data. In the context of drilling data, where numerous interrelated parameters exist, such techniques are useful for isolating the key drivers of performance metrics.

Time-series decomposition is another advanced preprocessing technique applied to separate observed data into trend, seasonal, and residual components. This decomposition facilitates the identification of cyclical patterns or irregularities that may impact drilling performance. By analyzing these components separately, models can be trained to account for seasonality or long-term trends, leading to more accurate and reliable predictions.

Incorporating domain expertise is vital throughout the preprocessing and feature engineering stages. Understanding the physical and operational context of each parameter allows for the creation of features that are not only statistically significant but also practically meaningful. For instance, integrating geological information, such as lithology logs, with mechanical parameters can yield features that reflect the interaction between equipment and specific rock formations, providing deeper insights into drilling dynamics.

3.2.4 Handling Missing Data and Noise in Drilling Time Series

Drilling time series datasets, such as those derived from the FORGE project, are inherently noisy and frequently contain missing data points. These issues arise from a variety of sources, including sensor malfunctions, communication errors, downhole environmental disturbances, and temporary data acquisition failures. In predictive modeling, these inconsistencies can introduce biases, reduce model accuracy, and undermine the reliability of drilling parameter forecasts. Therefore, effective strategies for handling missing data and noise are essential to ensure robust and accurate ML models for drilling optimization.

Missing data in drilling datasets typically manifest as gaps in time series records, often occurring intermittently and non-uniformly across different operational phases. For example, during critical drilling operations such as tripping or casing installation, certain sensors may be temporarily deactivated, resulting in irregular or incomplete time series segments. Addressing these gaps requires an understanding of both the statistical properties of the data and the operational context of the drilling process. Common approaches to handling missing data include interpolation, imputation, and predictive modeling techniques. Linear interpolation, for instance, estimates missing values based on adjacent data points, preserving temporal continuity. More advanced methods, such as spline interpolation, polynomial fitting, or Gaussian processes, can capture more intricate temporal patterns in the data. However, these methods are most effective when missing data points are sparse and follow predictable trends.

In scenarios where missing data patterns are more complex or involve larger gaps, ML-based imputation techniques, such as k-nearest neighbors or multiple imputations by chained equations, can offer more reliable solutions. These approaches leverage the relationships between different drilling parameters to estimate missing values in one variable based on observed values in others. For example, missing WOB measurements might be inferred from torque and rotational speed using regression-based imputation models. Such approaches have been effectively applied in prior studies of geothermal drilling datasets, where the interdependencies between operational parameters facilitated accurate imputation of missing time series data points.

Noise in drilling datasets is another challenge, often introduced by sensor inaccuracies, mechanical vibrations, or external environmental factors. Noise is particularly problematic in high-frequency measurements, where minor sensor inconsistencies can accumulate over time and distort long-term trends. Techniques for noise reduction commonly include smoothing filters, such as moving average filters, Savitzky-Golay filters, and wavelet denoising. Moving average filters are effective at reducing high-frequency noise by averaging data points over a defined window, but they may also smooth out sharp transitions that represent meaningful operational events. Savitzky-Golay filters, on the other hand, apply polynomial fitting within the moving window, preserving local features while reducing noise. Wavelet denoising offers a more sophisticated approach by decomposing the signal into multiple frequency bands and selectively filtering out high-frequency noise components without distorting the underlying data structure.

In addition to filtering techniques, statistical anomaly detection methods are often employed to identify and address outliers, which are frequently associated with sensor noise or transient operational disturbances. Methods such as z-score analysis and robust principal component analysis are particularly useful in identifying extreme values that deviate from the expected range of operational parameters. Outlier removal or replacement strategies must be carefully applied to ensure that meaningful anomalies, such as abrupt changes due to sudden geological transitions, are not inadvertently discarded during preprocessing.

Temporal inconsistencies, such as irregular sampling intervals, further exacerbate the challenges posed by noise and missing data. In drilling operations, data acquisition rates often vary based on operational phases, with higher sampling frequencies during critical activities such as bit trips or directional drilling. Standardizing time intervals across the dataset is crucial for downstream ML model training and evaluation. Techniques such as time resampling are frequently used to address these inconsistencies. For example, a uniform time grid can be established, and data points can be resampled using interpolation or aggregation methods, depending on the required temporal resolution. Careful consideration must be given to selecting appropriate resampling intervals to balance temporal granularity with computational efficiency.

Another advanced approach to handling noise and missing data involves data augmentation techniques, which are designed to artificially increase the dataset size while addressing its deficiencies. Time warping, noise injection, and window slicing are commonly used methods in this context. Time warping involves subtle transformations in the temporal alignment of data points, effectively simulating variations in drilling speed or operational timing. Noise injection introduces controlled perturbations to data points, enabling the model to learn robustness against sensor inconsistencies. Window slicing involves segmenting time series data into smaller overlapping windows, enhancing the model's ability to generalize across different temporal patterns.

Moreover, domain knowledge plays a role in distinguishing between genuine data anomalies and meaningful operational signals. For example, sudden pressure spikes in downhole measurements might signify equipment failure or abrupt geological transitions rather than sensor noise. Integrating contextual metadata, such as lithology logs and operator annotations, can enhance the preprocessing pipeline by providing additional validation layers for missing data imputation and noise detection.

3.2.5 Limitations and Generalizability

Although the Utah FORGE dataset offers high temporal resolution and rich instrumentation, it represents a single geothermal field setting with site specific geology, operational practices, and tool configurations. The experiments in this chapter include both single well analysis focused on the pilot well 58-32 and multiwell experiments conducted using additional wells from the same FORGE site. These design choices support controlled comparisons under consistent sensing and operational conditions and allow assessment of transfer within a single geothermal project. However, because all wells originate from the same site, the reported results do not establish cross site generalizability. Accordingly, performance should be interpreted as evidence of within well generalization for the pilot well and within site generalization across FORGE wells, rather than a guarantee of transferability to other geothermal fields, other lithologies, or different rig and sensor configurations. A more comprehensive assessment of generalizability would require evaluation on independent wells outside the FORGE project, ideally spanning different geothermal settings, drilling assemblies, and data acquisition systems. In addition, future work should include systematic uncertainty calibration checks across held out wells to determine whether prediction intervals remain reliable under operational and geological variation.

3.3 Transformer Model Design for ROP Prediction

3.3.1 Transformer Architecture Overview

The transformer architecture, introduced by Vaswani et al. (2017), has advanced sequence modeling by overcoming the limitations of traditional RNNs. Its self-attention mechanism enables the model to capture dependencies across input sequences without relying on sequential processing, making it particularly effective for time series forecasting tasks, such as predicting drilling parameters like ROP.

A standard transformer comprises an encoder-decoder structure (Fig. 3.2). The encoder processes the input sequence to generate an intermediate representation that encapsulates contextual relationships between elements. The decoder utilizes this representation, along with previously predicted outputs, to produce forecasts. Within both the encoder and decoder, two key components enhance the model's capabilities: multi-head self-attention and position-wise feed-forward networks. Multi-head self-attention allows the transformer to focus on different parts of the input sequence concurrently, effectively capturing both short-term and long-term dependencies. Each attention head independently computes attention scores, and their aggregation forms a comprehensive understanding of the sequence.

The self-attention mechanism operates by computing a weighted sum of input features, with weights determined by the similarity between query, key, and value vectors derived from the input data. This process enables the model to dynamically identify which parts of the input sequence are most relevant for a given prediction task, such as understanding the influence of WOB or RPM on the ROP.

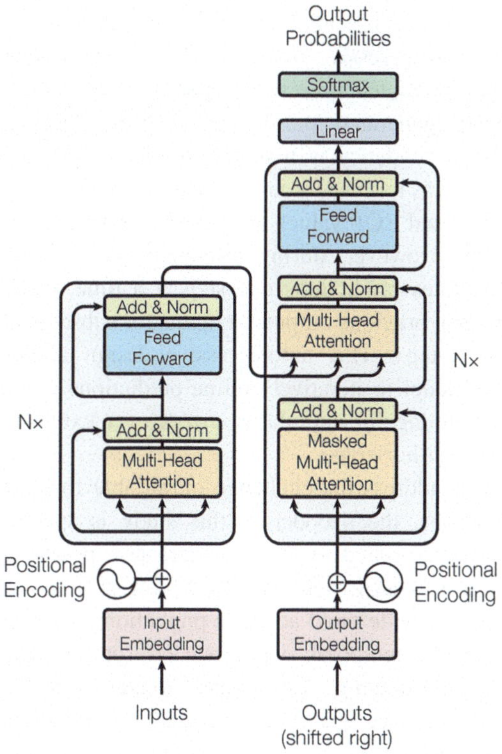

Fig. 3.2 Attention is all you need transformer architecture (Vaswani et al. 2017)

Position-wise feed-forward networks introduce non-linearity into the transformer model. Each layer of the encoder and decoder includes a fully connected feed-forward sublayer that applies transformations independently to each position in the sequence. These networks consist of two linear transformations with a ReLU activation function in between, enabling the model to capture complex relationships within the data.

One challenge in applying transformers to time series data is their lack of temporal awareness. To address this, transformers incorporate positional encodings, which inject information about the order of input data points directly into the model. These encodings ensure that the model retains information about the sequence order, crucial for accurate predictions in drilling operations where temporal trends affect performance metrics like MSE and drilling efficiency.

The scalability of transformers further contributes to their suitability for time series data in drilling workflows. By leveraging parallel processing, transformers can efficiently handle large volumes of sensor data collected from drilling operations. This capability is particularly relevant when processing real-time measurements from downhole tools, such as WOB, torque, and downhole pressure.

Recent advancements in transformer architectures have enhanced their ability to handle time series data effectively. For instance, the Autoformer model, introduced by Wu et al. (2021), incorporates a decomposition architecture with an auto-correlation mechanism, enabling it to capture intricate temporal patterns and model long-range dependencies more effectively.

Despite their strengths, transformers have limitations. Computational latency can be a concern in real-time drilling applications, where timely predictions are needed. Strategies such as model optimization and efficient inference architectures are being explored to address these constraints. Additionally, transformers require substantial datasets for effective training, and their performance can be affected by noisy or incomplete data. Therefore, thorough data preprocessing and feature engineering are required when preparing drilling datasets for transformer models.

In this research, transformer architectures are tailored to predict key drilling performance metrics, focusing on optimizing the attention mechanism, fine-tuning positional encodings, and incorporating domain-specific features derived from drilling parameters. By leveraging the capabilities of transformers, this study aims to develop a robust predictive framework capable of capturing both short-term operational variations and long-term geological trends, ultimately contributing to improved drilling efficiency and reduced operational costs.

3.3.1.1 Encoder-Decoder Framework for ROP Prediction

The encoder-decoder framework is the foundational architecture of transformer models and plays a central role in their effectiveness for time-series forecasting tasks such as predicting the ROP. Unlike traditional recurrent or convolutional models, the encoder-decoder structure enables transformers to simultaneously model short-term patterns and long-term dependencies within sequential data, addressing key challenges in drilling time series prediction.

In the context of ROP forecasting, the encoder processes historical drilling parameters, such as WOB, RPM, torque, and mud flow rate, to create a latent representation that captures the interdependencies and temporal relationships among these variables. Each parameter is first embedded into a high-dimensional space through learnable embeddings, which allow the model to represent raw numerical values in a form suitable for transformer layers. Positional encodings are then added to the embeddings to incorporate temporal order into the sequence. This is particularly important for drilling datasets, where the temporal arrangement of sensor readings directly influences predictive accuracy.

Within the encoder, the self-attention mechanism evaluates relationships between all time steps in the input sequence, assigning attention weights to each parameter based on their relevance to predicting future ROP values. For example, the encoder might learn that variations in WOB have a stronger impact on ROP under specific geological conditions, while torque plays a more prominent role in others. Multi-head self-attention enhances this process by allowing the model to consider multiple perspectives simultaneously, capturing distinct patterns and interactions among parameters across different temporal scales.

Once the encoder produces its intermediate representation, it passes this information to the decoder. The decoder is responsible for generating the predicted ROP sequence over a defined forecast horizon. Like the encoder, the decoder uses self-attention mechanisms but introduces an additional component known as encoder-decoder attention. This mechanism allows the decoder to attend to specific time steps from the encoder's output while generating predictions sequentially. Encoder-decoder attention helps the model dynamically focus on critical historical patterns that have the most influence on future ROP trends.

Another essential aspect of the decoder is its autoregressive nature, which means that each predicted time step depends on the previously generated predictions. During training, this dependency is managed through a process known as teacher forcing, where the model is fed the actual ROP values instead of its own predictions. However, during inference, the decoder generates predictions one step at a time, using its own previous outputs as inputs for subsequent time steps. This autoregressive design allows the model to iteratively refine predictions while accounting for cumulative errors over extended forecast horizons.

To address the challenge of variable forecast horizons, the decoder in this study is specifically fine-tuned to balance short-term precision with long-term stability. Short forecast horizons often provide more accurate predictions but may lack the temporal depth needed for strategic decision-making. Conversely, longer horizons offer broader operational insights but are more susceptible to error accumulation. By carefully selecting an optimal forecast horizon, informed by experimental results, the decoder is calibrated to deliver reliable predictions that align with the requirements of drilling workflows.

Feature engineering also plays a role in the encoder-decoder framework for ROP prediction. Derived metrics, such as MSE, are integrated as additional inputs to the transformer model. These engineered features provide the encoder and decoder with enriched contextual information, enabling more accurate representation and prediction of complex, nonlinear relationships within drilling data.

The encoder-decoder framework also benefits from advanced regularization techniques to prevent overfitting, particularly given the variability and noise present in drilling datasets. Techniques such as dropout, layer normalization, and gradient clipping are applied during training to ensure model stability and generalization across diverse operational conditions.

A computational challenge in the deployment of encoder-decoder architectures is managing latency during real-time predictions. In operational drilling workflows, predictions must be delivered with minimal delays to enable timely adjustments to drilling parameters. To address this, optimizations such as model pruning, quantization, and parallelized computation are integrated into the deployment pipeline. These optimizations reduce computational overhead without compromising predictive accuracy, ensuring that the encoder-decoder model can function reliably in live operational environments.

The encoder-decoder framework's ability to simultaneously model temporal dependencies and capture complex parameter interactions positions it as a highly effective architecture for ROP prediction. Its flexibility allows for adaptation to diverse drilling environments, from high-temperature geothermal wells to deepwater offshore projects. Moreover, the modular design of transformers enables iterative refinements and fine-tuning based on specific operational datasets, improving model performance over time.

3.3.1.2 Hyperparameter Tuning and Optimization Strategies

The performance of transformer-based models in predicting the ROP depends on effective hyperparameter tuning and optimization strategies. Transformers are highly flexible architectures with numerous hyperparameters, including attention heads, layer depth, dropout rates, batch sizes, and learning rate schedules. Fine-tuning these parameters is essential for achieving optimal model performance, especially when applied to complex, nonlinear drilling datasets. This section goes into key hyperparameter choices and optimization methodologies, drawing insights from existing research on transformer models for time-series forecasting.

One of the most important hyperparameters in transformer architectures is the number of attention heads within the multi-head self-attention mechanism. Attention heads allow the model to focus on different aspects of input data, capturing fine-grained dependencies between drilling parameters over various temporal horizons. Studies suggest that increasing the number of attention heads enhances the model's ability to process diverse temporal patterns, but it also raises computational overhead (Vaswani et al. 2017). For time-series forecasting tasks, a common practice is to use between four and eight attention heads, balancing performance and efficiency. However, the optimal number of heads depends on the dataset's temporal complexity and variability in operational parameters.

Another essential hyperparameter is the dropout rate, which serves as a regularization technique to prevent overfitting during model training. Dropout randomly disables a fraction of neurons in the transformer layers during training, reducing the risk of the model memorizing noise in the dataset. Research by Srivastava et al. (2014) suggests that dropout rates between 0.1 and 0.3 are generally effective for DL architectures. For transformers applied to drilling datasets, fine-tuning dropout rates is particularly important due to the sparse and noisy nature of the data.

Learning rate optimization also plays a central role in transformer training, as an improperly configured learning rate can lead to slow convergence or unstable training dynamics. Transformers often benefit from warm-up strategies during the early stages of training, where the learning rate gradually increases before stabilizing. Vaswani et al. (2017) introduced a learning rate scheduler with a warm-up phase followed by inverse square root decay, a widely adopted strategy for stabilizing transformer training. This approach allows the model to avoid premature convergence and effectively explore the loss surface during initial training epochs.

The choice of batch size is another crucial factor influencing transformer training efficiency. Larger batch sizes allow for parallel computation and stable gradient estimates but require substantial memory resources. However, extremely large batch sizes can sometimes cause training instability, necessitating batch size optimization tailored to the hardware and dataset constraints.

Hyperparameter search techniques are important for identifying the optimal configuration of transformer parameters. Traditional grid search approaches are computationally expensive and often infeasible for models as complex as transformers. Instead, random search and Bayesian optimization methods have gained traction for hyperparameter tuning. Bergstra and Bengio (2012) showed that random search often outperforms grid search by exploring a broader range of hyperparameter combinations in fewer iterations. Bayesian optimization techniques further refine this process by building probabilistic models to predict the most promising hyperparameter configurations based on previous evaluations.

Validation strategies are equally important in hyperparameter optimization, particularly for time-series data. Standard cross-validation techniques, which assume data independence, are not suitable for sequential datasets like drilling logs. Instead, rolling-origin cross-validation and expanding-window validation are commonly used to evaluate time-series models (Hyndman 2018). Rolling-origin validation iteratively shifts the training window forward in time while evaluating performance on subsequent time steps. Expanding-window validation progressively increases the training set size with each iteration, offering insights into how the model generalizes to unseen future data. These strategies are particularly effective for drilling datasets, where temporal dependencies play a role in predictive performance.

Lastly, early stopping is an essential strategy to prevent overfitting in transformer training. This technique monitors the model's performance on the validation set and halts training if performance stagnates or deteriorates over a defined number of epochs. Combined with checkpointing, where the best-performing model snapshot is saved during training, early stopping ensures optimal generalization without unnecessary computational expense.

3.3.1.3 Training Methodology

The negative log-likelihood (NLL) was employed as the loss function for training the transformer model. NLL measures the alignment between the predicted probability distribution of the model and the actual observed values, effectively quantifying how well the model captures the uncertainty in its forecasts. It is defined as:

$$L(\theta) = -\sum_{i=1}^{n} logp(y_i|\hat{y_i}, \theta) \tag{3.1}$$

where y_i represents the actual observed value, $\hat{y_i}$ denotes the predicted value,

θ encompasses the parameters of the chosen probability distribution, and

$p(y_i|\hat{y_i}, \theta)$ corresponds to the probability density function of the predicted distribution evaluated at the observed value. The Student's t-distribution was selected as the output distribution, enabling the model to account for heavy-tailed behavior and uncertainties frequently observed in drilling dynamics data. The loss function penalizes predictions that assign low probability to the true observed values, encouraging the model to produce accurate probabilistic forecasts. This approach is effective in scenarios characterized by noisy, time-dependent sensor data, as it captures not only the central tendency of predictions but also their associated uncertainty. By minimizing the NLL, the transformer model improves both the reliability and interpretability of its predictions across varying operational conditions and geological formations.

The symmetric mean absolute percentage error (SMAPE) was used as the primary evaluation metric. The SMAPE is a normalized metric that quantifies the accuracy of forecasts by measuring the percentage difference between predicted and actual values. It is defined as:

$$SMAPE = \frac{100\%}{N}\sum_{t=1}^{N}\frac{2 \times |F_t - A_t|}{|A_t| + |F_t|} \tag{3.2}$$

where F_t represents the forecasted value, A_t denotes the actual value, and

N is the number of observations. The symmetric nature of SMAPE addresses the asymmetry present in traditional percentage errors, providing a balanced perspective on overestimations and underestimations. SMAPE is effective in evaluating prediction accuracy in terms of relative percentage error rather than absolute deviations. This is particularly useful in drilling applications, where key parameters such as ROP vary across drilling depths and formations. By normalizing prediction errors relative to observed values, SMAPE balances the evaluation of performance across different operating conditions and reduces the influence of extreme outliers.

The Adam optimizer was used to minimize the negative log likelihood training objective associated with the selected output distribution. Convergence was monitored using the training objective during fitting, while SMAPE was computed on validation windows for model selection and reporting.

For this research, the dataset was partitioned using a time based strategy that preserved chronological order and aligned with the forecasting objective. Hyperparameters were selected using time based three fold cross validation so that each validation window occurred after its corresponding training window. After selection, the final model was refit on the designated training range and evaluated on a held out future segment reserved for testing. This approach respects temporal dependence and prevents leakage across splits, providing an evaluation that reflects how the model would be used in forward looking drilling operations.

3.3.1.4 Model Validation Strategy and Uncertainty Quantification

Model validation was structured to prevent information leakage by enforcing strict temporal separation between training and evaluation windows. Hyperparameter search used time based three fold cross validation so that each candidate configuration was scored on validation periods that occur after the corresponding training history. This design aligns with deployment conditions in which forecasts are generated using only prior measurements. The final configuration was then evaluated using a time based train test split, and performance was summarized using SMAPE as the primary accuracy metric.

Uncertainty quantification was implemented through probabilistic forecasting rather than point prediction. The transformer was trained using a negative log likelihood objective with Student's t as the output distribution to represent heavy tailed residuals and intermittent sensor noise observed in drilling time series. Under this formulation, each forecast step corresponds to a predictive distribution, enabling interval style summaries through forecast quantiles. Quantile outputs from 0.1 through 0.9 provide predictive bounds around the median forecast, while SMAPE remains the metric used to compare forecast accuracy across configurations and experiments.

Probabilistic outputs do not inherently guarantee calibrated interval coverage on unseen wells or operating regimes. Interval calibration should therefore be treated as a validation objective in future work, including empirical coverage testing and recalibration across held out wells within the same site and, when available, across independent geothermal sites.

The validation design includes both single well and multiwell experiments within the Utah FORGE dataset, but it does not establish generalizability beyond the FORGE site. The dataset reflects a controlled geothermal field setting with site specific operational practices, sensor behavior, and formation conditions. Although multiple wells are included, they are colocated within the same geothermal project. Results should therefore be interpreted as evidence of within well performance for the pilot well and within site transfer across forge wells, rather than guaranteed transfer to other geothermal fields. Future work should emphasize evaluation on independent wells outside the forge project when available, together with transfer learning studies and interval calibration tests that quantify whether uncertainty estimates remain reliable under distribution shift.

3.3.2 Experimental Results and Analysis

3.3.2.1 Sampling Interval Impact on Prediction Accuracy

In time series forecasting, particularly for applications such as predicting the ROP in drilling operations, the selection of an appropriate sampling interval is key. The sampling interval determines the frequency at which data points are recorded, directly influencing the model's ability to capture temporal patterns and its predictive accuracy.

In this study, we investigated the impact of various sampling intervals on the performance of a transformer-based DL model tasked with forecasting ROP. The objective was to identify the sampling interval that minimizes forecasting errors while preserving the patterns of the ROP measurements. We resampled the dataset at intervals ranging from 1 to 3600 s, maintaining a constant forecast horizon of 10 min. The SMAPE was employed as the evaluation metric to assess prediction accuracy (Fig. 3.3).

Our findings indicate a relationship between sampling interval length and model performance. As the sampling interval decreased from 3600 to 60 s, there was a notable improvement in the transformer's forecasting accuracy, evidenced by a reduction in SMAPE values. This enhancement can be attributed to the model's increased access to detailed temporal information, enabling it to capture short-term dynamics more effectively. However, when the sampling interval was further reduced to 1 s, a decline in performance was observed. This deterioration is likely due to the introduction of noise and less relevant fluctuations in high-frequency data, which can obscure meaningful patterns and adversely affect the model's learning process.

The optimal performance was achieved with a sampling interval of 60 s, suggesting that, for this dataset, this interval provides a balance between capturing essential temporal patterns and mitigating the effects of noise. It is important to note that while a 60-s interval was optimal for this specific dataset, the ideal sampling interval may vary depending on the characteristics of different datasets and the specific application requirements. Therefore, it is advisable to conduct a preliminary analysis to determine the most suitable sampling interval for each unique scenario.

3.3.2.2 Forecast Horizon Optimization for Multi-well or Single Well Comparison

In time series forecasting, particularly within the context of drilling operations, the forecast horizon, or the length of time into the future for which predictions are made, plays a key role in determining the accuracy and utility of predictive models. Selecting an appropriate forecast horizon is essential for capturing the temporal dynamics of the ROP and ensuring that predictions are both precise and actionable for well-specific operations.

In this study, we conducted a comprehensive analysis to optimize the forecast horizon for a transformer-based DL model aimed at predicting ROP (Fig. 3.4). The analysis encompassed two primary scenarios: first, evaluating a transformer model trained on data from multiple wells without any fine-tuning for the target well; and second, assessing a transformer model that underwent retraining using data specific to the target well to enhance its adaptability

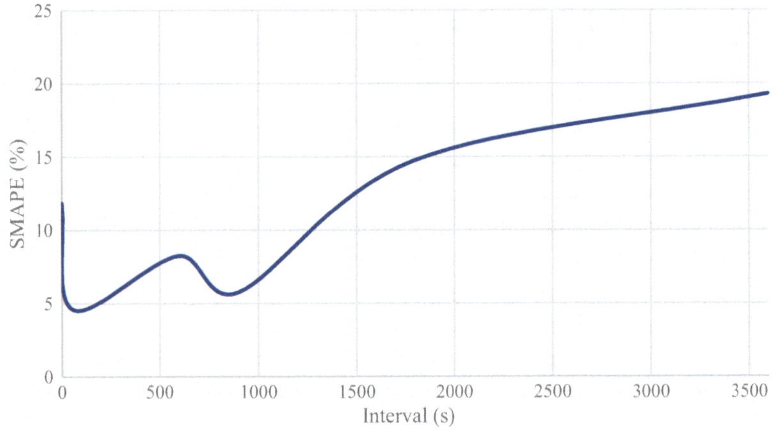

Fig. 3.3 Sampling interval impact on ROP forecaster performance (Urdaneta et al. 2024)

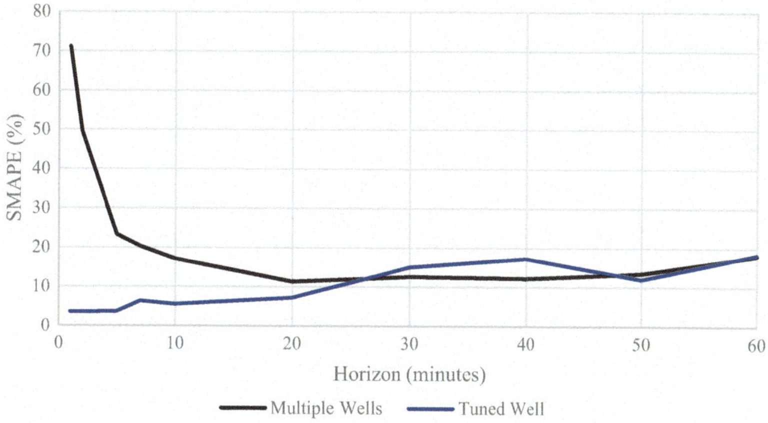

Fig. 3.4 Impact of forecast horizon tuning on performance for multiple wells, where the Y-axis represents SMAPE, and the X-axis denotes the forecast horizon time in minutes (Urdaneta et al. 2024)

to unique geological and operational conditions.

For the multi-well transformer model, our experiments revealed that a forecast horizon of 20 min yielded the lowest prediction error, with a SMAPE of 11.3%. This finding suggests that, in the absence of well-specific retraining, the model can capture general patterns across different wells up to a 20-min horizon. However, the performance at this horizon may not fully account for the unique characteristics of individual wells, potentially limiting its predictive accuracy in well-specific applications.

To address this limitation, we implemented a retraining strategy wherein the transformer model was fine-tuned using a subset of data from the target well. This approach aimed to tailor the model to the specific geological formations and operational parameters of the well in question. The retrained, well-specific transformer model demonstrated superior performance compared to the multi-well model, particularly at shorter forecast horizons. Notably, a forecast horizon of 10 min was identified as a pragmatic balance, achieving a lower SMAPE and providing sufficient lead time for drillers to make informed decisions regarding parameter adjustments.

It is important to recognize that while shorter forecast horizons can enhance predictive accuracy, they may also impose practical constraints on operational decision-making. Drillers require adequate time to interpret predictions and implement necessary adjustments to drilling parameters. Therefore, the selection of an optimal forecast horizon must consider both the model's predictive performance and the operational feasibility of acting upon the forecasts.

Furthermore, the variability of geological formations and drilling conditions across different wells requires a flexible approach to forecast horizon selection. In wells where geological conditions change rapidly, shorter forecast horizons may be more effective in capturing dynamic shifts, whereas in more stable environments, longer horizons could suffice. Consequently, conducting a preliminary analysis of the specific well conditions is advisable to determine the most suitable forecast horizon for each application.

3.3.2.3 Transformer Forecaster Performance Across Well Sections

In this section, we go into the performance evaluation of the transformer-based forecaster across various sections of the FORGE pilot well 58–32, focusing on its ability to predict the ROP over a 10-min forecast horizon. The analysis encompasses depth intervals at 6792–6793, 7140–7142, 7511–7514, 7517–7519, 7530–7532, and 7534–7536 ft, providing a comprehensive assessment of the model's predictive capabilities in different depth contexts.

Figures 3.5, 3.6, 3.7, 3.8, 3.9, and 3.10 illustrate the comparative plots of actual versus predicted ROP values, alongside corresponding

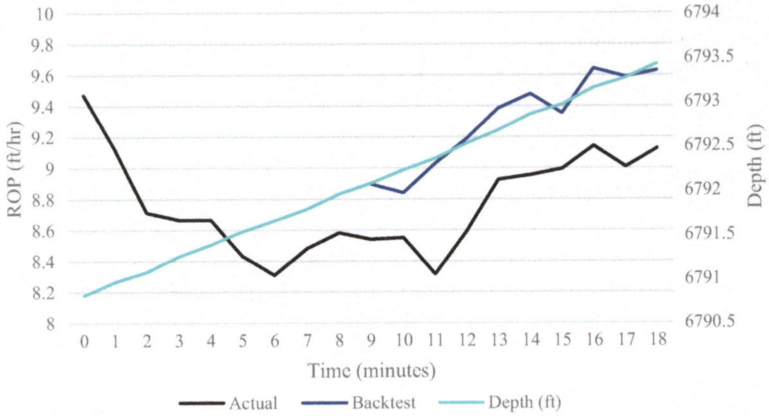

Fig. 3.5 ROP forecaster prediction and depth versus time for pilot well FORGE 58-32 for 6792–6793 ft section (Urdaneta et al. 2024)

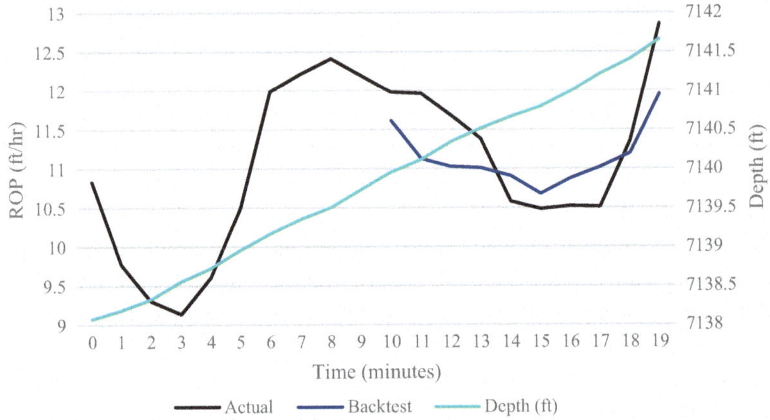

Fig. 3.6 ROP forecaster prediction and depth versus time for pilot well FORGE 58-32 for 7140–7142 ft section (Urdaneta et al. 2024)

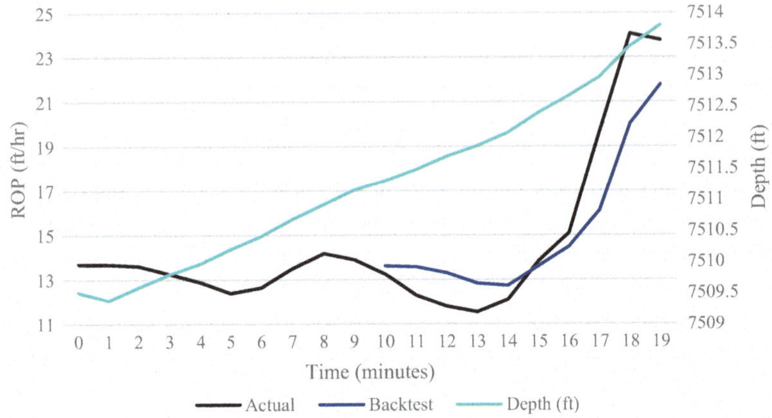

Fig. 3.7 ROP forecaster prediction and depth versus time for pilot well FORGE 58-32 for 7511–7514 ft section (Urdaneta et al. 2024)

depth profiles over time. In these visualizations, the black line represents the actual ROP measurements, serving as the ground truth benchmark. The blue line denotes the model's backtested predictions, reflecting its performance during periods with known outcomes. Additionally, the aqua line indicates the well depth during the prediction intervals, offering contextual insight into the drilling progression.

Across the evaluated depth sections, the transformer forecaster achieved an average SMAPE of 5.2%, compared with 11.9% for DeepAR

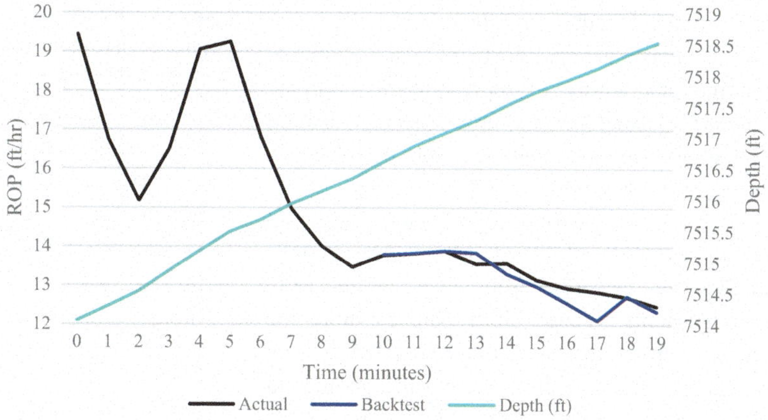

Fig. 3.8 ROP forecaster prediction and depth versus time for pilot well FORGE 58-32 for 7517–7519 ft section (Urdaneta et al. 2024)

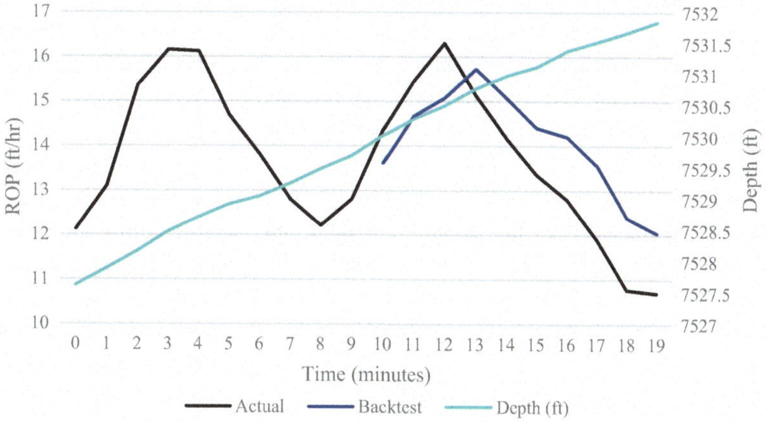

Fig. 3.9 ROP forecaster prediction and depth versus time for pilot well FORGE 58-32 for 7530–7532 ft section (Urdaneta et al. 2024)

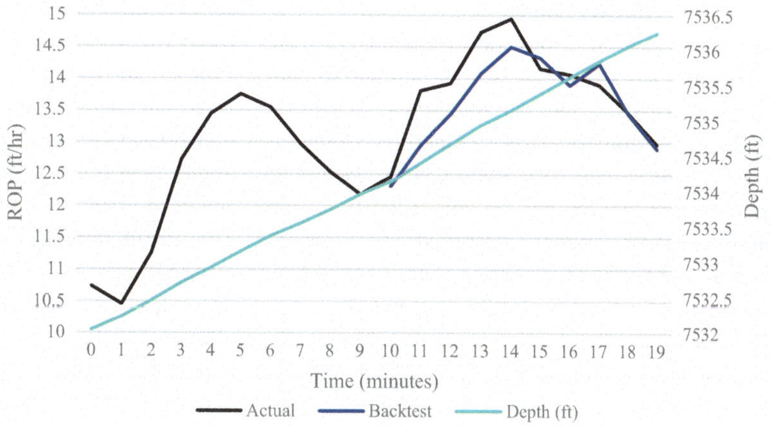

Fig. 3.10 ROP forecaster prediction and depth versus time for pilot well FORGE 58-32 for 7534–7536 ft section (Urdaneta et al. 2024)

and 11.2% for MQ-CNN, corresponding to an approximate 53–56% reduction in prediction error relative to the alternative architectures. In four of the six evaluated intervals, the transformer recorded the lowest SMAPE among the three models, with values as low as 2.2% and 2.6% in the 7517–7519 ft and 7534–7536 ft sections, respectively. These results indicate that the transformer was able to capture the dominant temporal trends and operational variations in the ROP data across multiple geological conditions. The alignment between predicted and measured

ROP values, together with the consistently lower error metrics across most intervals, supports the model's ability to represent the complex temporal dependencies characteristic of drilling operations while maintaining stable performance across varying depths.

Despite the strong performance, certain anomalies were observed where the model's predictions deviated from the actual ROP measurements. These discrepancies may be attributed to factors such as unmodeled drilling dynamics, sensor noise, or abrupt geological transitions that are not adequately captured in the training data. Such anomalies highlight the challenges in predictive modeling within complex and variable subsurface environments.

It is important to acknowledge the limitations associated with the sample size used for validation. The reported accuracy is based on the model's validation on specific segments of well data. To enhance both the quantity of data and the quality of results, we suggest validating the model on larger datasets derived from diverse wells and operational contexts as future work.

3.3.2.4 Comparative Evaluation with Alternative Architectures

In this section, the performance of the transformer-based forecaster is compared with two alternative DL architectures, DeepAR and the Multi-Horizon Quantile Convolutional Neural Network (MQ-CNN). DeepAR, introduced by Salinas et al. (2020) is an autoregressive model designed for probabilistic time-series forecasting, while MQ-CNN, introduced by Wen et al. (2017) employs convolutional layers to capture temporal dependencies across multiple forecast horizons. Both models have demonstrated success in industrial time-series applications, but their performance in capturing the unique dynamics of CTD remains underexplored.

The objective of this comparison is to evaluate the predictive capabilities of each model in forecasting the ROP across several distinct depth sections of the FORGE pilot well 58-32. The assessment focuses on key evaluation metrics, including SMAPE, mean absolute scaled error, and root mean squared error, to provide a quantitative measure of prediction accuracy and consistency.

Table 3.1 Transformer test scores across different sections of the FORGE pilot well 58-32 compared to other DL architectures (Urdaneta et al. 2024)

Depth (ft)	Transformer SMAPE (%)	DeepAR SMAPE (%)	MQ-CNN SMAPE (%)
6792–6793	5.5	2.7	2.8
7140–7142	4.2	5.3	2.8
7511–7514	9.1	20.2	16.8
7517–7519	2.2	21.0	6.5
7530–7532	7.7	11.2	18.7
7534–7536	2.6	10.9	19.5
Average	5.2	11.9	11.2

The results, summarized in Table 3.1, illustrate the comparative performance of the three architectures across six specific well depth intervals. These intervals were chosen to capture a range of geological and operational conditions, ensuring a representative evaluation of each model's adaptability. Across all evaluated sections, the transformer forecaster demonstrated superior predictive accuracy, achieving an average SMAPE of 5.22%, outperforming DeepAR, which recorded an average SMAPE of 11.9%, and MQ-CNN, which recorded 11.2%.

At the depth interval of 6792–6793 ft, the transformer recorded a SMAPE of 5.5%, notably higher than both DeepAR and MQ-CNN, which achieved errors of 2.7% and 2.8%, respectively. This result indicates that, in this interval, both DeepAR and MQ-CNN captured the temporal dependencies and operational patterns more effectively than the transformer. A possible explanation could be the localized nature of the patterns in this depth range, where recurrent and convolutional architectures may have been better suited to identify short-term dependencies.

At the 7140–7142-foot interval, the transformer improved its relative performance, achieving a SMAPE of 4.2%. DeepAR, in comparison, recorded an error of 5.3%, while MQ-CNN outperformed the transformer's performance with an error of 2.8%. In this scenario, the transformer outperformed DeepAR but underperformed comparably to MQ-CNN, suggesting that both MQ-CNN most effectively captured the temporal dependencies in this depth section.

A more pronounced difference is observed in the 7511–7514-foot section, where the transformer recorded a SMAPE of 9.1%. While this value reflects relatively higher prediction error compared to other intervals, it is notably better than DeepAR's error of 20.2% and MQ-CNN's error of 16.8%. These results suggest that the transformer architecture was better equipped to handle the increased variability and nonlinear dependencies encountered in this depth section, outperforming the recurrent and convolutional models despite its higher absolute error compared to other intervals.

At the 7517–7519-foot interval, the transformer demonstrated a clear advantage with an exceptionally low SMAPE of 2.2%, outperforming DeepAR and MQ-CNN, which recorded errors of 21.0% and 6.5%, respectively. This result shows the transformer model's capacity to maintain accuracy in variable operational conditions, effectively capturing both short-term fluctuations and long-term dependencies.

In the 7530–7532-foot section, the transformer recorded a SMAPE of 7.7%, which was lower than DeepAR's 11.2% and MQ-CNN's 18.7%. Finally, in the 7534–7536-foot interval, the transformer achieved a SMAPE of 2.6%, outperforming DeepAR, which recorded an error of 10.9%, and MQ-CNN, which recorded 19.5%. This interval reinforces the consistency of the transformer in achieving lower error values across diverse geological and operational conditions.

The observed variations in performance across depth intervals can be attributed to differences in how each architecture processes temporal dependencies and noise in the data. DeepAR, as an autoregressive recurrent neural network, is better suited for tasks with strong sequential patterns but struggles with long-term dependencies and noisy, irregular time-series data. Its relatively poor performance in several intervals suggests sensitivity to these challenges. MQ-CNN, while effective in capturing localized temporal dependencies, appears to struggle with modeling long-range dependencies and interactions between drilling parameters, resulting in inconsistent performance across depth intervals.

In contrast, the transformer architecture excels in managing both short-term and long-term dependencies through its self-attention mechanism. This mechanism allows the model to dynamically focus on relevant time steps without being constrained by sequential processing, making it particularly well-suited for drilling datasets characterized by irregular sampling, noise, and multi-scale dependencies. Additionally, the parallel processing capabilities of transformers enable them to maintain stable performance across varying time horizons and depth intervals, reducing the risk of cumulative error propagation.

However, it is important to acknowledge that the transformer did not consistently outperform the other architectures across all evaluated intervals. At certain localized depths, such as 6792–6793 ft, both DeepAR and MQ-CNN demonstrated lower SMAPE values. These discrepancies may be explained by localized noise, abrupt changes in operational conditions, or insufficient training examples for those specific geological formations. Addressing these limitations through well-specific retraining or enhanced feature engineering could further improve the transformer's adaptability.

3.3.3 Conclusion

The comparative analysis of transformer-based predictive models for drilling performance optimization, specifically targeting the ROP, has demonstrated advancements in accuracy, adaptability, and robustness across varying operational and geological conditions. This chapter examined the performance of transformer architectures through a series of experiments, including assessments of sampling interval impact, forecast horizon optimization, retraining strategies, performance across multiple well sections, and comparative evaluation with alternative DL architectures. The findings collectively show the potential of transformer models in addressing challenges associated with real-time drilling parameter prediction.

The analysis began with an investigation into the influence of sampling intervals on predictive accuracy. It was established that sampling interval selection plays a role in balancing data granularity and model stability. The results revealed that shorter intervals, while theoretically providing more detailed temporal information, can introduce noise and unnecessary computational overhead, degrading prediction accuracy. Conversely, excessively long sampling intervals risk omitting operational dynamics. An optimal sampling interval of 60 s was identified, striking a balance between temporal resolution and noise minimization. This interval allowed the transformer architecture to focus on meaningful recurring patterns in the data while mitigating the effects of noise and inconsistencies encountered in drilling datasets.

The subsequent analysis focused on forecast horizon optimization, exploring the relationship between prediction timeframes and model performance. It was observed that shorter forecast horizons often yield higher prediction accuracy, given the reduced propagation of errors over time. However, excessively short horizons may not provide sufficient lead time for decision-making in live drilling environments. The results indicated that a forecast horizon of ten minutes achieved an optimal balance, providing both accuracy and practical utility. Additionally, the transformer demonstrated notable adaptability across different forecast horizons, further validating its suitability for dynamic drilling environments where operational conditions can shift rapidly.

Retraining strategies were also explored to evaluate the adaptability of the transformer model to well-specific conditions. The analysis revealed that retraining with localized data improved predictive accuracy, allowing the model to adjust its parameters to reflect unique geological and operational characteristics. The retrained transformer consistently outperformed baseline models trained on generalized datasets, particularly in depth sections characterized by abrupt geological transitions or operational anomalies. These findings emphasize the importance of incorporating well-specific retraining workflows into practical drilling optimization frameworks to ensure sustained model performance across diverse operational contexts.

Performance evaluation across multiple well sections provided further insights into the transformer's capabilities. The model demonstrated consistent accuracy across varying depth intervals, effectively capturing both short-term variations and long-term trends in ROP data. While minor performance inconsistencies were observed in specific depth ranges, the transformer generally exhibited resilience to noise, temporal irregularities, and changing geological conditions.

The comparative evaluation against alternative architectures, including DeepAR and MQ-CNN, further highlighted the transformer's advantages. Across most evaluated well sections, the transformer achieved lower error rates, demonstrating superior accuracy in predicting ROP trends. However, the evaluation also highlighted certain limitations. The transformer did not uniformly outperform the alternative architectures across all evaluated sections. In isolated intervals, both DeepAR and MQ-CNN exhibited lower error rates. These findings suggest that further refinements in preprocessing workflows, including noise reduction and feature engineering, may enhance the transformer's adaptability to outlier conditions. Additionally, the computational overhead associated with transformer architectures remains a concern for real-time deployment, particularly in resource-constrained environments. Future research should prioritize optimization strategies, such as model pruning, to address these challenges.

References

Abbaszadeh Shahri, A., Shan, C., Larsson, S., & Johansson, F. (2024). Normalizing large scale sensor based MWD data: An automated method toward a unified database. Sensors, 24(4), 1209.

Aoun, M. A. B., & Madarász, T. (2022). Applying machine learning to predict the rate of penetration for geothermal drilling located in the Utah FORGE site. Energies, 15(12), 4288.

Atashnezhad, A., Akhtarmanesh, S., Hareland, G., & Al Dushaishi, M. (2021). Developing a drilling optimization system for improved overall rate of penetration in geothermal wells. In ARMA US Rock Mechanics/ Geomechanics Symposium. ARMA.

Bergstra, J., & Bengio, Y. (2012). Random search for hyper-parameter optimization. Journal of Machine Learning Research, 13, 281–305.

Biggio, L., Bendinelli, T., Kulkarni, C., & Fink, O. (2023). Ageing-aware battery discharge prediction with deep learning. Applied Energy, 346, 121229.

Ding, Z., Chen, Y., Wang, H., Wang, X., Zhang, W., & Zhang, Y. (2025). TIformer: A transformer-based framework for time-series forecasting with missing data. In Australasian Database Conference (pp. 71–84). Springer.

Gidh, Y., Purwanto, A., & Smith Bits. (2012). Artificial neural network drilling parameter optimization system improves ROP by predicting and managing bit wear. In SPE Intelligent Energy International Conference and Exhibition. SPE.

Girase, H., Agarwal, N., Choi, C., & Mangalam, K. (2023). Latency matters: Real-time action forecasting transformer. In Proceedings of the IEEE/CVF Conference on Computer Vision and Pattern Recognition (pp. 18759–18769).

Gwynn, M., Allis, R., Hardwick, C., Jones, C., Nielsen, P., & Hurlbut, W. (2018). Rock properties of FORGE Well 58–32, Milford, Utah. Transactions, Geothermal Resources Council.

Hegde, C., Soares, C., & Gray, K. (2018). Rate of penetration modeling using hybrid models: Deterministic and machine learning. In Unconventional Resources Technology Conference (pp. 3220–3238).

Hyndman, R. J. (2018). Forecasting: Principles and practice. OTexts.

Iglesias, G., Talavera, E., González-Prieto, Á., Mozo, A., & Gómez-Canaval, S. (2023). Data augmentation techniques in time series domain: A survey and taxonomy. Neural Computing and Applications, 35(14), 10123–10145.

Islam, N., Vijapurapu, R., Jones, M., McLennan, J., Moore, J., Rickard, W., Balamir, O., & Vagnetti, R. (2018). Application of mechanical specific energy and at-the-bit measurements for geothermal drilling applications. In ARMA US Rock Mechanics/ Geomechanics Symposium. ARMA.

Motahhari, H. R., Hareland, G., & James, J. A. (2010). Improved drilling efficiency technique using integrated PDM and PDC bit parameters. Journal of Canadian Petroleum Technology, 49(10), 45–52.

Phelan, Z., Xing, P., Panja, P., Moore, J., & McLennan, J. (2022). Prediction of formation properties based on drilling data of wells at Utah FORGE site using machine learning. In ARMA US Rock Mechanics/ Geomechanics Symposium. ARMA.

Rickard, W. M., McLennan, J., Islam, N., & Rivas, E. (2019). Mechanical specific energy analysis of the FORGE Utah well. In Proceedings of the 44th Workshop on Geothermal Reservoir Engineering. Stanford University.

Salinas, D., Flunkert, V., Gasthaus, J., & Januschowski, T. (2020). DeepAR: Probabilistic forecasting with autoregressive recurrent networks. International Journal of Forecasting, 36(3), 1181–1191.

Srivastava, N., Hinton, G., Krizhevsky, A., Sutskever, I., & Salakhutdinov, R. (2014). Dropout: A simple way to prevent neural networks from overfitting. Journal of Machine Learning Research, 15, 1929–1958.

Urdaneta, C., Jeong, C., Wu, X., & Chen, J. (2024). Deep learning method for improving rate of penetration prediction in drilling. SPE Journal.

Vaswani, A., Shazeer, N., Parmar, N., Uszkoreit, J., Jones, L., Gomez, A. N., Kaiser, Ł., & Polosukhin, I. (2017). Attention is all you need. Advances in Neural Information Processing Systems.

Wen, R., Torkkola, K., Narayanaswamy, B., & Madeka, D. (2017). A multi-horizon quantile recurrent forecaster. arXiv preprint arXiv:1711.11053.

Wu, H., Xu, J., Wang, J., & Long, M. (2021). Autoformer: Decomposition transformers with auto-correlation for long-term series forecasting. Advances in Neural Information Processing Systems, 34, 22419–22430.

Zhou, H., Zhang, S., Peng, J., Zhang, S., Li, J., Xiong, H., & Zhang, W. (2021). Informer: Beyond efficient transformer for long sequence time-series forecasting. In Proceedings of the AAAI Conference on Artificial Intelligence (Vol. 35, pp. 11106–11115).

Coiled-Tubing Drilling Dynamics Forecasting

4

4.1 Background and Related Work

CTD represents an advancement in the drilling industry, offering improved mobility, reduced environmental impact, and enhanced operational flexibility compared to conventional rotary drilling systems. Its ability to enable underbalanced drilling, where the wellbore pressure is kept lower than the formation pressure, has been instrumental in preventing formation damage and improving reservoir productivity. However, despite these advantages, drilling dynamics in coiled tubing operations remain inherently complex, with challenges arising from the interplay of operational parameters, downhole shock, and ROP.

Downhole shock remains among the most important challenges in CTD. Excessive shock levels can cause rapid equipment wear, damage downhole tools, and lead to NPT. A key contributor to elevated shock levels in CTD is the use of underbalanced drilling techniques, where wellbore pressure is intentionally maintained lower than the formation pressure. While this approach effectively prevents formation damage and enhances productivity, it increases susceptibility to downhole vibrations and shock. The pressure differential creates an environment where fluid dynamics and mechanical interactions between the drill bit and formation become more erratic, amplifying oscillations along the drill string. Additionally, the continuous spool of coiled tubing, unlike the segmented rigid pipes used in conventional drilling, introduces unique mechanical behaviors, including a higher propensity for helical buckling which alters the dynamics of the drill string, increasing the likelihood of vibrational instabilities. Together, these factors introduce nonlinearities into the drilling system, making it more challenging to predict and mitigate shock-related dysfunctions. These complexities require predictive models capable of capturing both short-term and long-term trends in CTD drilling dynamics to enable proactive operational adjustments.

Accurately predicting ROP and downhole shock is critical for drilling performance and operational safety. While ROP serves as a fundamental indicator of drilling efficiency, directly affecting the time and cost of operations, downhole shock levels influence tool integrity, bit life, and the risk of mechanical failures. Optimizing these metrics is far from straightforward due to the nonlinear dependencies among drilling parameters such as WOB, rotational speed, drilling fluid properties, and formation characteristics. Traditional empirical and physics-based models have attempted to address these relationships but often fall short when faced with the highly dynamic and variable conditions encountered in real-world drilling operations.

ML techniques have increasingly been applied to address these limitations, offering a data-driven approach to capturing complex relationships within drilling datasets. Previous

C. Urdaneta et al., *Deep Learning Applications for Drilling Performance: Forecasting, Diagnostics, Telemetry, and Predictive Maintenance*, Advances in Oil and Gas Exploration & Production,
https://doi.org/10.1007/978-3-032-26648-4_4

studies have demonstrated the efficacy of ML models, including random forests, support vector machines, and artificial neural networks, in predicting ROP and detecting downhole dysfunctions. For example, Aoun and Madarász (2022) successfully applied ML algorithms to predict ROP for geothermal drilling at the Utah FORGE site, achieving improvements over traditional regression models. Similarly, Alsaihati and Alotaibi (2022) employed ML techniques to predict the severity of downhole vibrations, highlighting the advantages of data-driven approaches in managing drilling dynamics.

While these models have demonstrated promise, they are often constrained by their limited ability to capture long-term dependencies within time-series data. RNNs and LSTM architectures, while effective in modeling sequential dependencies, are computationally intensive and suffer from vanishing gradient problems when dealing with long sequences of data. Transformers, originally introduced for natural language processing, have emerged as a powerful alternative for time-series forecasting due to their self-attention mechanism. This architecture allows transformers to weigh the relative importance of each data point in a sequence, enabling them to model both short-term variations and long-range dependencies with remarkable efficiency (Wang et al. 2023).

Recent advancements in transformer-based models have highlighted their potential in predicting drilling dynamics. Transformers excel at identifying subtle temporal patterns in time-series data, making them particularly suitable for applications where real-time responsiveness and predictive accuracy are important. Wang et al. (2023) demonstrated the superiority of transformer architectures over traditional time-series models in anomaly detection tasks, showing their ability to handle noisy and incomplete datasets.

This study focuses on applying a transformer-based architecture specifically designed for analyzing CTD datasets. The dataset used includes real-world operational data encompassing parameters such as WOB, downhole pressure, and flowrate. The objective is to predict both ROP and downhole shock with high accuracy, enabling real-time decision-making and proactive mitigation of drilling dysfunctions.

The results presented in this study highlight the strengths and limitations of each architecture, providing insights into their suitability for drilling dynamics prediction. Furthermore, the integration of domain-specific preprocessing techniques, feature engineering, and parameter optimization ensures that the models are tailored to address the unique challenges posed by CTD operations.

4.1.1 Challenges in Multi-parameter Prediction for Drilling Dynamics

The prediction of drilling dynamics, focusing on the ROP and shock behavior, presents challenges due to the complex and interdependent nature of drilling parameters. In CTD, these complexities are amplified by the unique characteristics of the dataset, including high temporal variability, nonlinear parameter interactions, and the dynamic operational adjustments required to optimize both ROP and mitigate excessive shock.

One of the primary challenges lies in capturing the relationships between key drilling parameters. Variables such as WOB, flowrate, mud flow rate, and downhole torque exhibit strong interdependencies, often governed by geological and mechanical conditions. For example, increasing WOB can improve penetration rates under some drilling conditions, but it may result in accelerated bit wear or induce excessive vibration in other conditions. These nonlinear and context-dependent interactions require predictive models capable of adapting to shifting conditions while preserving accuracy.

Drilling datasets, including the CTD data used in this study, are characterized by both short-term oscillations and long-term trends. Transient spikes in shock levels can indicate mechanical dysfunctions or sudden geological transitions, while gradual declines in penetration rates may suggest bit wear or changing rock properties. Models must effectively distinguish between noise, transient anomalies, and meaningful long-term trends to provide reliable predictions. The real-time nature of the CTD dataset, with frequent sampling intervals capturing these temporal patterns, adds both richness and complexity to the predictive task.

A unique challenge in this study arises from the dual focus on predicting both ROP and shock behavior. Unlike single-task predictive frameworks, which focus on a single output variable, this multi-task approach requires capturing the interdependencies between these two metrics. Operational changes aimed at optimizing penetration rates, such as increasing rotational speed or adjusting mud flow rates, can inadvertently amplify shock levels. Similarly, interventions to minimize shock may come at the expense of reduced drilling efficiency. Balancing these competing objectives within a single predictive framework introduces an additional layer of complexity, demanding architectures that can effectively model both outputs without sacrificing accuracy in either domain.

The geological variability in the CTD dataset introduces another layer of unpredictability. Drilling operations frequently encounter transitions between formations with varying rock strength, abrasiveness, and porosity. For instance, moving from soft, porous formations to hard, abrasive rock can cause abrupt changes in torque, shock behavior, and penetration rates. Predictive models must remain adaptable across these transitions without requiring retraining for each geological shift. This adaptability is particularly relevant in real-world operational settings, where time and computational constraints limit the feasibility of frequent retraining cycles.

Preprocessing the CTD dataset also experienced challenges, as noise, sensor anomalies, and missing values are common in drilling data. Sensor inaccuracies, intermittent malfunctions, and environmental factors often introduce artifacts that obscure meaningful patterns. Additionally, inconsistent sampling intervals can create temporal misalignments, further complicating model training. In this study, preprocessing steps, including anomaly detection, noise filtering, and temporal alignment, were employed to address these issues.

The comparative evaluation of different predictive architectures, including transformer models, DeepAR, and MQ-CNN, introduces an additional layer of complexity to predicting drilling dynamics. Each architecture has distinct design principles that influence its strengths and limitations in handling time-series forecasting tasks. Transformer models are known for their ability to capture long-term temporal dependencies and handle large-scale datasets effectively through self-attention mechanisms (Vaswani et al. 2017). In contrast, DeepAR, introduced by Salinas et al. (2020), employs an autoregressive recurrent neural network structure and has been applied successfully to time-series tasks where sequential patterns and probabilistic forecasting play a role. Meanwhile, MQ-CNN, proposed by Wen et al. (2017), integrates convolutional layers with a quantile decoder to predict future values across different horizons, which can be effective for modeling complex temporal patterns in datasets with varying resolution.

However, while these architectures have demonstrated strengths in other domains, their application to drilling dynamics prediction, particularly with CTD datasets that include both ROP and shock behavior, remains underexplored. This chapter specifically evaluates how these architectures handle the challenges presented by the CTD dataset, including parameter interdependencies, temporal variability, and dual-task prediction requirements. By analyzing their predictive performance under these conditions, this study aims to provide clarity on their relative strengths and weaknesses in the context of drilling dynamics forecasting.

4.2 Dataset Description and Preparation

4.2.1 CTD Dataset Overview and Key Parameters

The CTD dataset analyzed in this study allows to investigate the dynamic interactions between drilling parameters, ROP, and downhole shock behavior. This dataset originates from operational wells where CTD techniques were employed, capturing real-time measurements from downhole and surface sensors across varying geological and operational conditions. The dataset encompasses multiple drilling phases, including wiper trips, lateral drilling, and directional adjustments, providing a diverse representation of both routine operations and anomalous events. This diversity ensures that the dataset captures the variability in CTD dynamics, including transitions between different lithological formations, equipment responses to parameter adjustments, and the temporal evolution of drilling performance metrics.

The dataset consists of time series data collected from an array of sensors installed both along the coiled tubing MWD tool and at the surface. These sensors record a range of drilling parameters, including WOB, downhole torque, mud flow rate, annular pressure, and downhole vibrations. These parameters are complemented by depth measurements and metadata detailing lithology logs, drilling tool configurations, and operational annotations recorded by field engineers. Together, these data streams create a multidimensional representation of drilling operations, capturing both mechanical and environmental influences on ROP and shock behavior.

Among the key parameters included in the dataset, WOB serves as a primary driver of drilling efficiency. It represents the axial force applied to the drill bit and directly influences cutting efficiency and penetration rates. Optimal WOB settings vary depending on lithology, bit design, and real-time operational conditions. Excessive axial force can lead to bit wear, stick–slip vibrations, and inefficient energy transfer, while insufficient force can result in poor rock-breaking efficiency and reduced penetration rates.

Rotational speed, another important parameter, reflects the rate at which the drill bit rotates and contributes to rock fragmentation and drilling efficiency. In CTD, unlike conventional drilling, the coiled tubing pipe itself does not rotate. Instead, rotational energy is provided by either a downhole motor or a turbine powered by the hydraulic energy of drilling fluids. Flow rate plays a key role in transferring this energy, as the circulating fluid drives the motor or turbine, which in turn rotates the drill bit. The rotational speed of the bit is thus influenced by both the differential pressure across the motor or turbine and the flow rate of the drilling fluid. These parameters act as primary indicators of the bit's rotational efficiency and mechanical performance. However, the relationship between rotational speed, WOB, and ROP is non-linear and heavily dependent on formation type, bit design, and the operational efficiency of the downhole motor or turbine. Incremental increases in flow rate or WOB do not always result in proportional improvements in penetration rates. Mechanical dysfunctions, such as torque stalls, motor inefficiencies, or hydraulic energy losses, can further disrupt the expected relationships between these parameters. Additionally, transient conditions like rapid pressure fluctuations or inconsistent fluid delivery can introduce variability in rotational speed, impacting both ROP and drilling performance.

Downhole torque measurements provide insights into the resistance encountered by the drill bit as it interacts with the rock face. Torque serves as an indirect indicator of drilling efficiency and mechanical stability, with abrupt fluctuations often signaling dysfunctions such as differential sticking or borehole instability. Additionally, torque data exhibit temporal patterns that correspond to varying lithological properties and changes in drilling parameters.

Mud flow rate is another essential parameter, as drilling fluids play a role in cooling the drill bit, transporting cuttings to the surface, and maintaining wellbore pressure stability. Flow rate variations can affect downhole pressure gradients, hole cleaning efficiency, and MSE. Insufficient flow rates may result in poor cuttings removal and increased vibration intensity, while excessive flow rates can induce turbulence and energy losses, ultimately impacting drilling efficiency.

Downhole vibrations, including axial, lateral, and torsional vibrations, are among the most important factors influencing drilling dynamics in CTD operations. Unlike conventional drilling setups, where drill pipes offer more rigid structural responses, the continuous and flexible nature of coiled tubing introduces additional mechanical complexities. Helical buckling and underbalance drilling can cause dynamic instabilities that propagate along the tubing string. These instabilities exacerbate vibration severity and contribute to uneven wear on drilling tools. Vibration data captured in the dataset reflect these interactions, providing valuable insights into shock propagation and mechanical dysfunctions that affect both penetration rates and equipment reliability.

Depth measurements serve as a spatial reference for correlating drilling parameters with specific geological formations. Depth-specific trends in ROP and downhole shock provide insights into how lithology affects drilling performance. For example, transitions from soft, porous formations to hard, abrasive rock might correspond to abrupt shifts in ROP, torque, and vibration intensity.

The metadata accompanying the dataset further enrich its analytical value. Detailed lithology logs provide contextual information about subsurface rock properties, including compressive strength, abrasiveness, porosity, and permeability. Operational annotations recorded by field engineers offer qualitative insights into key events, such as equipment failures, parameter adjustments, and drilling strategy changes. These metadata elements serve as auxiliary information, enabling better interpretation of time series patterns.

The CTD dataset used in this study was collected during field operations conducted by SLB using a coiled tubing MWD system. The data represent operational measurements acquired under commercial drilling programs and are confidential. As a result, the raw dataset cannot be released publicly. However, the preprocessing workflow, feature definitions, and evaluation methodology are described in this chapter to allow reproduction of the modeling approach using comparable datasets.

4.2.2 Data Preprocessing and Feature Engineering

The preprocessing and feature engineering of the CTD dataset ensures the accuracy of subsequent predictive modeling. Drilling datasets, especially those collected in real-world operational conditions, are often characterized by irregular sampling intervals, sensor noise, missing values, and parameter inconsistencies. These issues must be addressed to prepare the dataset for effective modeling and to ensure that important temporal dependencies and parameter interactions are preserved.

Handling missing data is a fundamental aspect, as sensor malfunctions, communication delays, and transient disruptions result in incomplete records. In this study, interpolation techniques were employed to estimate missing values within short gaps, leveraging adjacent time steps to reconstruct plausible values. For more extensive data gaps, imputation strategies relying on parameter correlations were used, where related variables, such as flow rate and downhole pressure, informed the estimation of missing values. Careful consideration was given to preserving temporal consistency during this process, ensuring that imputed data did not distort time-series patterns.

Noise reduction was another key step, given the sensitivity of downhole sensors to environmental conditions, mechanical vibrations, and equipment interactions. Noise can obscure meaningful trends and introduce false patterns into the dataset, undermining model accuracy. Smoothing filters, including moving averages and Savitzky Golay filters, were applied to reduce noise while retaining essential temporal dynamics. Additionally, statistical anomaly detection techniques, such as z-score analysis, were used to identify and handle extreme outliers that could affect the learning process.

Temporal resampling was applied to address inconsistencies in sampling intervals, which are common in drilling datasets due to variations in data acquisition frequencies across different sensors and systems. Standardizing the time intervals across all parameters ensured temporal uniformity, which is essential for training predictive models that rely on consistent input–output sequences.

Feature normalization played an important role in mitigating the effects of varying parameter scales. Drilling parameters, such as flow rate, differential pressure, and WOB, operate across different numerical ranges. Without normalization, parameters with larger absolute values could affect the model's learning process. Min–max normalization was applied to scale the parameters within a defined range, ensuring that all features contributed equitably to the predictive framework.

Feature engineering extended beyond basic preprocessing to derive new variables that capture meaningful relationships among drilling parameters. Pairwise linear and polynomial combinations of key input features were explored to uncover dependencies within the dataset. These engineered features enabled the model to better capture non-linear interactions and complex dependencies between parameters, enhancing its ability to generalize across varying operational conditions. Variable importance analysis was also employed to identify which features contributed most to the predictive outcomes, ensuring that important insights were preserved and emphasized in the model's learning process.

Finally, the dataset was partitioned into training, validation, and testing subsets using a time-based approach that preserved the chronological order of drilling operations. The most recent segment of the data, aligned with the forecast horizon, was reserved as a hold-out test set to evaluate the model's ability to predict future drilling performance. The preceding portion was utilized for training, with additional time-based cross-validation employed to fine-tune hyperparameters and validate model stability across different temporal windows. This strategy ensured that the model learned from historical trends, optimized its parameters on sequential validation folds, and demonstrated generalization on unseen future data.

4.3 Transformer Model for ROP and Shock Dynamics in CTD

4.3.1 Methodology for Transformer-Based Prediction Models in CTD Dynamics

In this study, separate transformer models were developed for predicting ROP and shock behavior, rather than a single unified dual-output architecture. This choice was driven by the distinct statistical properties, temporal patterns, and operational significance of each target variable. While both ROP and shock share common influencing factors, such as WOB, flow rate, and differential pressure, their underlying behaviors differ. ROP typically exhibits smoother temporal trends, with gradual shifts influenced by mechanical efficiency, bit-rock interaction, and formation characteristics. In contrast, shock behavior displays abrupt, high-frequency fluctuations driven by transient mechanical interactions, including stick–slip events, torsional vibrations, and rapid pressure variations. These differences require distinct transformer architectures optimized for each variable to ensure precise predictive performance.

Both transformer models adopted an encoder-decoder architecture tailored to time-series forecasting. The encoder processed temporal sequences through stacked self-attention layers, enabling the model to capture both short-term fluctuations and long-term dependencies. Positional encodings were integrated to ensure that temporal order within the data was preserved, allowing each data point to be interpreted in the correct temporal context. The decoder utilized multi-head attention mechanisms to focus on the most relevant regions of the encoded sequence, transforming these representations into meaningful predictions for ROP and shock values. This architectural design was fine-tuned to address the operational complexities and dynamic interactions characteristic of CTD.

The training process for both models involved iterative optimization using historical datasets collected from operational drilling environments. The loss function employed during training was the negative log-likelihood, as detailed in Sect. 3.3.1.3. This loss function was selected due to its effectiveness in capturing probabilistic uncertainties and aligning predictions with the heavy-tailed distribution patterns commonly observed in drilling data. The Student's t-distribution was used as the output distribution, enhancing the model's ability to account for extreme values and variability in both ROP and shock predictions.

The Adam optimizer was employed to guide gradient descent during training, with dynamic learning rate scheduling implemented to adjust the learning rate based on training progress. Dropout regularization was applied within attention layers and feed-forward networks to mitigate overfitting, while early stopping mechanisms ensured that training ceased once validation loss stabilized. These strategies collectively contributed to efficient convergence and robust generalization across unseen data.

Model performance was evaluated using the SMAPE, also defined in Sect. 3.3.1.3. SMAPE provided a balanced metric for assessing prediction accuracy across varying magnitudes of drilling parameter values. Validation was performed using a separate dataset that was not included during training, enabling an unbiased assessment of model performance. Additionally, backtesting was conducted to compare historical predictions against actual observed outcomes, further validating the models' reliability in capturing temporal dynamics.

4.3.2 Model Validation Strategy and Uncertainty Quantification

Model validation was designed to prevent information leakage by enforcing strict temporal separation between training history and evaluation windows. Hyperparameter search used time based three fold cross validation, where each candidate configuration was evaluated on validation periods that occur after the corresponding training segment. This design approximates a deployment setting in which forecasts are generated using only past measurements. Final evaluation used the selected configuration and summarized accuracy using SMAPE for both rate of penetration and downhole shock, with reported values computed over the held out evaluation windows used in the chapter experiments.

Uncertainty quantification was handled through probabilistic forecasting rather than point prediction. The transformer was trained with a negative log likelihood objective and a Student's t output distribution to represent heavy tailed residuals and intermittent sensor noise that occur in drilling time series. Under this formulation, each forecast step corresponds to a predictive distribution, enabling interval style summaries through forecast quantiles. In practice, quantile outputs can be interpreted as predictive bounds around the central tendency, while SMAPE remains the external metric used to compare forecast accuracy across candidate configurations and across experiments.

Because probabilistic outputs do not guarantee calibrated coverage under distribution shift, interval calibration should be treated as a validation objective in future work. This includes coverage testing and recalibration across held out operational intervals and, when available, across held out wells from the same operator and tool configuration.

The evaluation in this chapter includes experiments across multiple well sections and operational regimes present in the CTD dataset, but it does not establish generalizability beyond the specific field context represented by the available data. Differences in bottom hole assembly, coiled tubing string properties, sensor packages, and operating practices can change both the dynamics and the residual structure that the model must learn. Therefore, results should be interpreted as evidence of forecasting performance within the studied CTD setting, rather than guaranteed transfer to other fields or operations without revalidation. Future work should emphasize evaluation on independent CTD datasets outside the studied field context when available, along with transfer learning experiments and interval calibration tests that quantify whether uncertainty estimates remain reliable under distribution shift.

4.4 Experimental Results and Analysis

4.4.1 Transformer Forecaster Performance Across Well Sections

This section evaluates the performance of the transformer-based architecture in predicting ROP and downhole shock across multiple well sections in CTD operations. The analysis was conducted on unseen real-world drilling data to ensure an unbiased assessment of the model's ability to generalize beyond the training dataset. The objective was to examine how effectively the transformer model captures the temporal evolution of these parameters under varying operational conditions, including trend reversals, sustained upward or downward trends, and sudden dynamic fluctuations.

The time-series plots illustrate the predicted and actual values for both ROP and downhole shock across selected well sections. Figures 4.1 and 4.2 depict the predictions for ROP across two representative intervals, showcasing the model's ability to follow the trends and transitions observed in the drilling data. In Fig. 4.1, the model demonstrates robust predictive alignment during an upward trend continuation phase. The forecasted values mirror the observed upward trajectory, with minor deviations during transitional phases where operational conditions shift rapidly. Similarly, Fig. 4.2 focuses on a downward trend reversal scenario, where the transformer model captures the deceleration phase accurately before predicting the subsequent rebound. While slight delays are evident during sharp inflection points, the trend alignment remains consistent with the observed data.

Figures 4.3 and 4.4 shift focus to the prediction of downhole shock. In Fig. 4.3, the transformer model accurately predicts an upward trend continuation in lateral shock. However, minor lag is observable during sudden accelerations in shock magnitude, reflecting the challenge of capturing instantaneous dynamic responses in complex drilling environments. In Fig. 4.4, the model handles a downward trend reversal scenario, aligning well with the

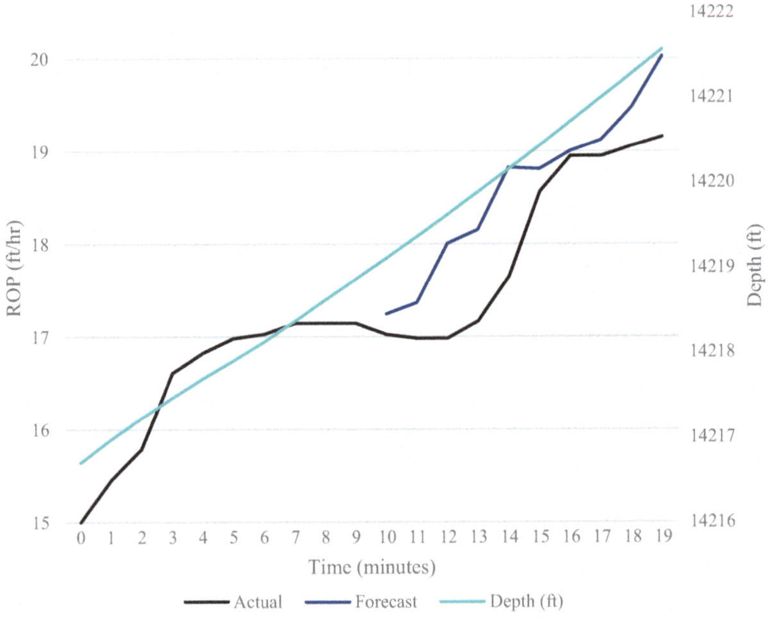

Fig. 4.1 ROP transformer prediction during upward trend continuation (Urdaneta et al. 2024)

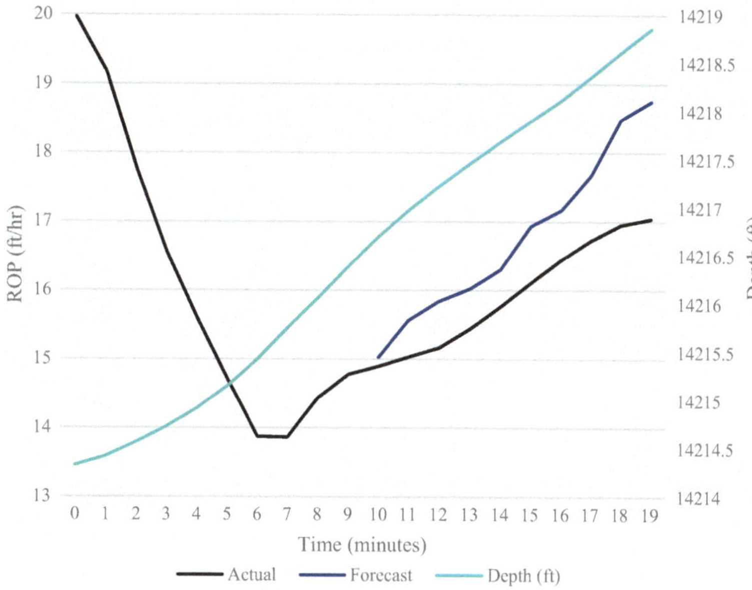

Fig. 4.2 ROP transformer prediction during downward trend reversal (Urdaneta et al. 2024)

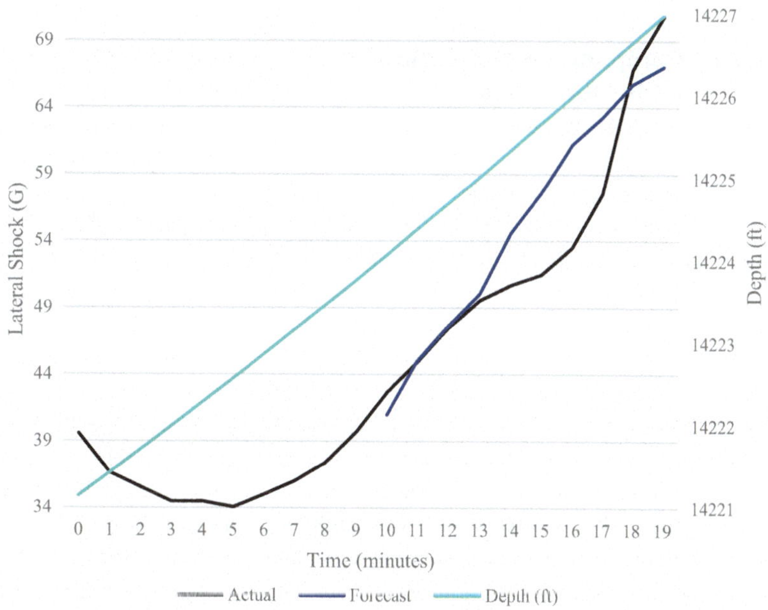

Fig. 4.3 Shock transformer prediction of upward trend continuation (Urdaneta et al. 2024)

observed trajectory during both the decline and subsequent recovery phases.

These results demonstrate the transformer architecture's strength in capturing both short-term fluctuations and long-term dependencies across different operational scenarios. The model's predictive accuracy remains consistent across varied well sections, effectively adapting to changes in drilling dynamics. Nevertheless, performance limitations emerge during periods of sharp transitions or outlier events, suggesting potential avenues for further optimization.

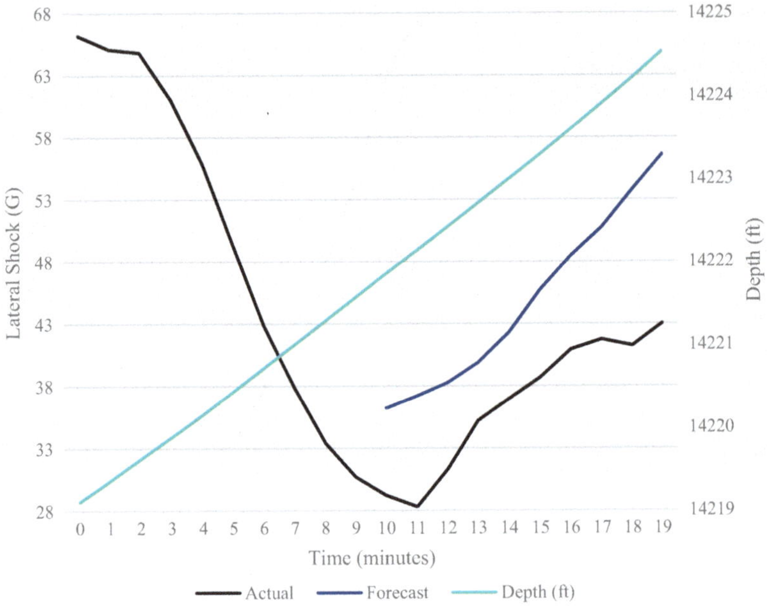

Fig. 4.4 Shock transformer prediction during downward trend reversal (Urdaneta et al. 2024)

4.4.2 Comparative Evaluation with Alternative Architectures

To comprehensively evaluate the performance of the transformer-based prediction model in forecasting CTD ROP and downhole shock dynamics, a comparative analysis was conducted against two commonly used time series forecasting architectures: DeepAR and MQ-CNN. All three models were trained and validated on an identical dataset comprising real-world CTD data. This consistent evaluation framework ensured that observed differences in performance could be attributed to the architectures themselves rather than variations in data preprocessing or model training strategies. The SMAPE was employed as the primary metric for comparison, offering a normalized evaluation of predictive accuracy across diverse operating conditions and drilling parameters.

The results, presented in Table 4.1, demonstrate differences in predictive performance across the three architectures. For shock predictions, the transformer model achieved an average SMAPE of 12.6%, while DeepAR performed slightly better with an average error of 11.0%. In contrast, MQ-CNN showed a higher average error of 30.4%, indicating limitations in its ability to capture the temporal dependencies and transitions associated with downhole shock dynamics. Specific intervals further illustrate these differences. In the depth range between 14,236 and 14,238 ft, MQ-CNN exhibited an error of 54.7%, whereas the transformer model and DeepAR achieved errors of 24.3% and 25.0%, respectively. Similarly, in the interval spanning 14,240 to 14,242 ft, MQ-CNN recorded an error of 50.6%, highlighting its struggle with high-frequency variations. In contrast, the transformer model demonstrated better adaptability with an error of 11.3%, though DeepAR performed marginally better with an error of 14.1%.

For ROP predictions, the transformer model achieved an average SMAPE of 12.8%, which was higher than DeepAR's average error of 6.6% but lower than MQ-CNN's average error of 15.9%. In certain intervals, DeepAR excelled in capturing smooth and consistent patterns, as seen in the depth range between 14,221 and 14,224 ft, where it achieved an error of 1.5% compared to the transformer model's 9.3% and MQ-CNN's 5.4%. However, in regions characterized by rapid transitions, such as 14,236 to 14,238 ft, the transformer model displayed

Table 4.1 Shock and ROP prediction test scores across different sections of a well (Urdaneta et al. 2024)

Depth (ft)	Shock Transformer SMAPE (%)	Shock DeepAR SMAPE (%)	Shock MQ-CNN SMAPE (%)	ROP Transformer SMAPE (%)	ROP DeepAR SMAPE (%)	ROP MQ-CNN SMAPE (%)
14,221–14,224	7.9	8.3	16.4	9.3	1.5	5.4
14,224–14,227	19.2	15.9	38.8	7.0	1.8	20.0
14,227–14,230	11.8	10.0	12.5	14.0	3.6	11.1
14,231–14,234	12.0	10.5	28.1	14.9	7.3	7.2
14,234–14,236	8.4	6.3	13.9	1.7	5.4	9.8
14,236–14,238	24.3	25.0	54.7	23.4	12.9	32.3
14,238–14,240	15.2	8.5	25.2	19.0	8.0	35.9
14,240–14,242	11.3	14.1	50.6	6.6	17.4	9.3
14,242–14,244	12.4	9.6	35.9	22.8	2.4	24.4
14,244–14,247	11.8	7.4	48.1	10.1	6.4	9.9
14,247–14,249	5.1	4.9	18.0	15.6	4.8	12.0
14,249–14,250	11.7	12.0	22.0	9.1	7.5	13.5
Average	12.6	11.0	30.4	12.8	6.6	15.9

better resilience with an error of 23.4%, whereas DeepAR recorded 12.9% and MQ-CNN reached 32.3%. This indicates the transformer model's relative strength in managing fluctuating conditions, even if its average performance was slightly less optimized than DeepAR in smoother operational ranges.

The comparative analysis reveals distinct strengths and weaknesses among the three architectures. DeepAR exhibited superior performance in predicting smoother, more stable temporal patterns, which is reflected in its consistently lower errors for ROP predictions in steady-state intervals. However, the transformer model demonstrated a balanced performance across both ROP and shock predictions, particularly in challenging regions where abrupt changes and high-frequency dynamics were prevalent. Its self-attention mechanism enabled it to dynamically allocate focus across relevant temporal dependencies, enhancing its robustness in regions with varying operational conditions.

MQ-CNN, while effective in isolated cases, struggled with maintaining accuracy across intervals with variability and abrupt transitions. Its higher average error rates for both shock and ROP predictions suggest limitations in its ability to model long-term dependencies and manage dynamic temporal patterns effectively.

4.5 Conclusion

This chapter presented a detailed investigation into the application of transformer-based models for predicting ROP and downhole shock dynamics in CTD operations, emphasizing their performance across multiple well sections and in comparison with alternative architectures such as DeepAR and MQ-CNN. The study leveraged real-world operational drilling data, ensuring the evaluation captured the complexities and nuances of field conditions rather than relying on idealized or simulated datasets. By focusing on key metrics and time-series visualizations, the analysis provided clarity on how transformer models handle the intricate temporal dependencies and abrupt dynamic changes in drilling environments.

The results demonstrated that transformer architectures excelled in modeling both short-term fluctuations and long-term trends within the drilling dataset, showcasing their adaptability across varying operational scenarios.

Specifically, the transformer model demonstrated strong alignment with observed data during periods of consistent operational behavior, such as sustained upward or downward trends, and maintained reasonable accuracy during sudden transitions and trend reversals. The time-series analyses revealed that while the transformer occasionally exhibited lag during rapid dynamic shifts, it consistently captured the trends in both ROP and shock values. This capability is important for drilling operations, where accurate predictions can guide real-time decisions, optimize drilling parameters, and mitigate the risk of equipment damage caused by excessive vibrations.

The comparative evaluation with DeepAR and MQ-CNN, summarized in Table 4.1, provides a quantitative view of the transformer model's performance across the evaluated intervals. While DeepAR achieved lower SMAPE values in certain localized sections, the transformer maintained a lower average error across the full evaluation set, indicating more consistent predictive accuracy under varying operational conditions. MQ-CNN showed competitive performance in some intervals, but its overall error metrics remained higher than those of the transformer in the majority of sections. These results suggest that the transformer provided more stable performance across diverse drilling conditions, while the alternative architectures were more sensitive to localized temporal patterns.

MQ-CNN, on the other hand, demonstrated weaker predictive performance across both ROP and shock predictions. Its error rates remained higher, particularly in intervals marked by frequent oscillations and sudden parameter shifts. This underperformance suggests that MQ-CNN struggled to capture the intricate temporal dependencies and non-linear interactions in the dataset, limiting its applicability to real-world drilling dynamics where abrupt changes are common. DeepAR excelled in stable conditions and the transformer model. These findings emphasize the importance of aligning model selection with the specific operational characteristics of the target drilling environment.

An important observation from this analysis was the ability of transformer models to allocate attention across relevant temporal regions, weighing the significance of recent and historical data points. This self-attention mechanism enabled the transformer to handle long-range dependencies and prioritize important temporal patterns without the memory constraints typically associated with traditional RNNs.

Despite these strengths, the study also identified areas where transformer models exhibited limitations. The occurrence of slight prediction lags during sharp transitions and sudden spikes in downhole shock suggests the need for further refinement. More importantly, the variability in model performance across different well sections shows the influence of geological variability and operational adjustments on predictive accuracy.

References

Alsaihati, A., & Alotaibi, B. (2022). Determining severity of lateral and torsional downhole vibrations while drilling surface holes using three machine learning techniques. SPE Journal, 27(3), 1493–1503.

Aoun, M. A. B., & Madarász, T. (2022). Applying machine learning to predict the rate of penetration for geothermal drilling located in the Utah FORGE site. Energies, 15(12), 4288.

Salinas, D., Flunkert, V., Gasthaus, J., & Januschowski, T. (2020). DeepAR: Probabilistic forecasting with autoregressive recurrent networks. International Journal of Forecasting, 36(3), 1181–1191.

Urdaneta, C., Jeong, C., Wu, X., & Chen, J. (2024). Predicting coiled-tubing drilling dynamics using transformers. In Unconventional Resources Technology Conference. URTeC.

Vaswani, A., Shazeer, N., Parmar, N., Uszkoreit, J., Jones, L., Gomez, A. N., Kaiser, Ł., & Polosukhin, I. (2017). Attention is all you need. Advances in Neural Information Processing Systems.

Wang, J., Sajeev, S. K., Ozbayoglu, E., Baldino, S., Liu, Y., & Jing, H. (2023). Reducing NPT using a novel approach to real-time drilling data analysis. In SPE Annual Technical Conference and Exhibition. SPE.

Wen, R., Torkkola, K., Narayanaswamy, B., & Madeka, D. (2017). A multi-horizon quantile recurrent forecaster. arXiv preprint arXiv:1711.11053.

5 Resistivity Tool Drilling Diagnostics

5.1 Background and Related Work

5.1.1 Importance of Resistivity Measurements in CTD Geosteering

Resistivity measurements play a foundational role in geosteering and formation evaluation, providing key insights into subsurface lithology and fluid saturation. In CTD operations, which are increasingly used for re-entry wells and multilateral developments, maintaining real-time awareness of formation properties is important for trajectory adjustments and optimal reservoir contact. Unlike conventional rotary drilling assemblies, CTD systems often operate in complex well architectures with rapid changes in inclination and azimuth, magnifying the importance of high-fidelity downhole measurements to support geosteering decisions.

Resistivity logging-while-drilling (LWD) wave propagation tools are designed to provide real-time estimations of formation resistivity by transmitting electromagnetic waves into surrounding rock and recording how these waves are attenuated and phase-shifted as they travel. These measurements are critical for differentiating between permeable hydrocarbon-bearing zones and non-productive intervals, directly informing trajectory adjustments to keep the wellbore within the target formation. As noted by Hermawan Manuaba et al. (2024), careful interpretation of electromagnetic propagation resistivity responses is fundamental for reliable formation evaluation, which ultimately influences the efficiency of well placement and the extent of reservoir contact.

Figure 5.1 illustrates how these LWD resistivity tools capture both attenuation and phase shift across formation boundaries. The plotted curves show characteristic responses as the tool moves from one lithological unit to another, providing a basis for interpreting changes in formation resistivity. In this example, the attenuation curve typically reflects the energy loss of the electromagnetic wave, while the phase shift indicates the delay introduced as the wavefront traverses different resistivity contrasts. These complementary measurements enable geoscientists and drilling engineers to identify bed boundaries, assess formation fluids, and steer the wellbore to maximize hydrocarbon exposure. The ability to interpret such data in real time is central to maintaining the well trajectory within the reservoir, enhancing hydrocarbon recovery, and supporting broader reservoir management strategies.

This is particularly important in CTD operations, where continuous circulation and the use of a non-rotating coiled tubing string limit the transmission of surface torque, making directional adjustments primarily dependent on downhole motors and controlled toolface orientation. Unlike conventional rotary drilling, which employs rotary steerable systems or slide drilling

C. Urdaneta et al., *Deep Learning Applications for Drilling Performance: Forecasting, Diagnostics, Telemetry, and Predictive Maintenance*, Advances in Oil and Gas Exploration & Production,
https://doi.org/10.1007/978-3-032-26648-4_5

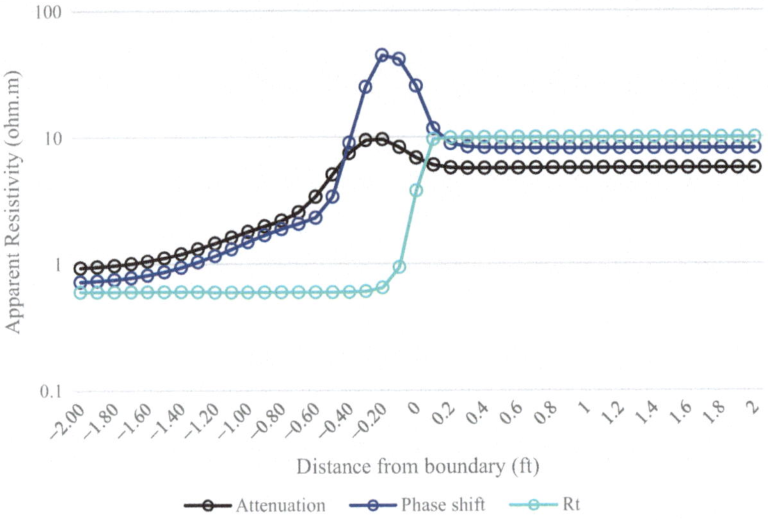

Fig. 5.1 Typical LWD wave-propagation tool response across a formation boundary. The black curve represents the apparent resistivity derived from signal attenuation, while the blue line represents the apparent resistivity derived from phase shift measurements. The teal line indicates the true formation resistivity R_t.. The horizontal axis shows distance from the formation boundary in feet, and the vertical axis shows apparent resistivity in ohm m (Hermawan Manuaba et al. 2024)

with drill pipe capable of transmitting substantial surface rotation and torque, CTD directional control relies on surface manipulation of the tubing and downhole hydraulic forces to maintain the desired trajectory. This means that real-time resistivity measurements play an even more important role by providing immediate formation feedback needed to keep the wellbore within target zones. Accurate data reduces the likelihood of undetected trajectory deviations that could later require more complex and mechanically sensitive corrective measures, such as orienting the coiled tubing string to reestablish toolface or executing costly trips to adjust the bottom-hole assembly (BHA).

Moreover, resistivity logs serve as a key input for updating subsurface models in real time. Integrated workflows often combine resistivity measurements with gamma-ray and azimuthal images to refine structural interpretations while drilling. In high-angle and horizontal wells, interpreting resistivity contrasts helps identify bed boundaries and ensures the wellbore remains within target hydrocarbon zones, reducing the likelihood of encountering unwanted water or gas intervals. This supports both immediate drilling decisions and longer-term reservoir management by minimizing the risk of early water or gas production.

5.1.2 Common Challenges in Resistivity Logging and Tool Failures

LWD wave propagation resistivity tools face several operational and interpretational challenges that can complicate their use, especially in complex geological settings. Figure 5.2 highlights key effects that influence measurement quality and reliability. Limited vertical resolution in thin layered formations can obscure bed delineation, while polarization effects near bed boundaries often skew readings, leading to misinterpretation of true formation resistivity. Proximity to adjacent beds with contrasting properties further complicates the resistivity response, introducing ambiguity into the log data.

Anisotropy within the formation adds another layer of complexity. It can distort the phase shift response more significantly than the amplitude,

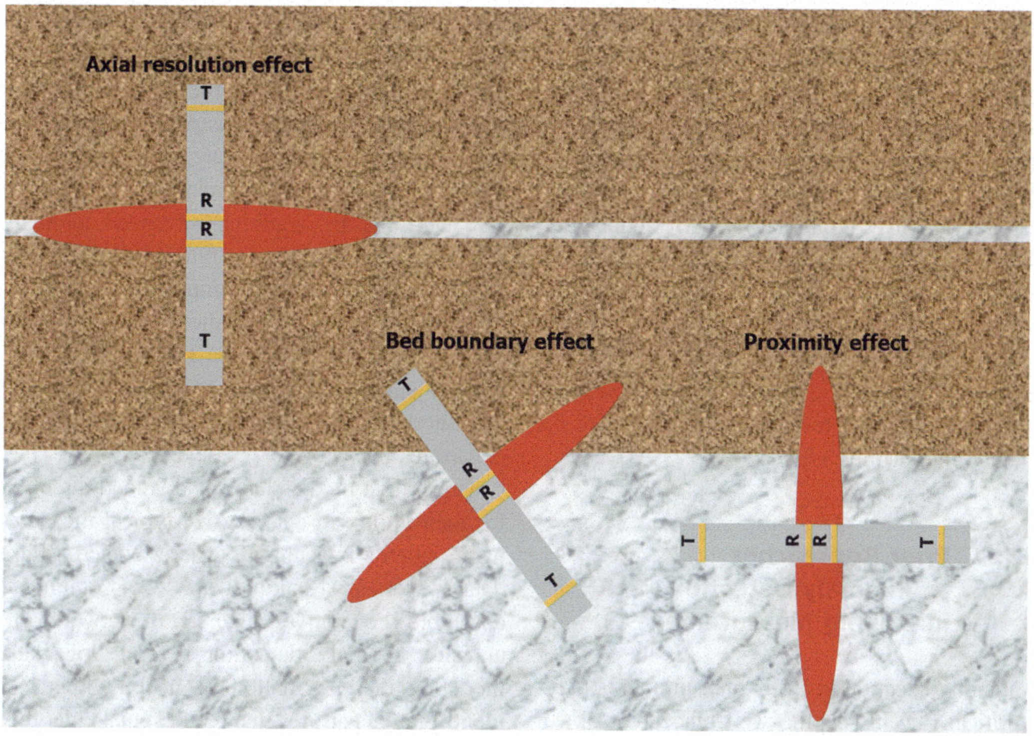

Fig. 5.2 Illustration of axial resolution, bed boundary, and proximity effects on LWD resistivity measurements (Hermawan Manuaba et al. 2024)

necessitating deeper interpretational corrections to differentiate between true resistivity variations and anisotropic effects. Additionally, amplitude measurements are generally less sensitive to anisotropy, which places greater emphasis on accurately understanding phase shift data. These challenges become even more pronounced in high-angle or horizontal well trajectories, where tool orientation relative to the formation layers exacerbates issues such as proximity effects and polarization. As Hermawan Manuaba et al. (2024) emphasize, differentiating between genuine resistivity contrasts and artifacts introduced by these phenomena is key for precise formation characterization and for making informed geosteering adjustments.

Despite their key role in guiding well placement, resistivity LWD tools face substantial operational and environmental challenges that can compromise data quality and tool lifespan. In CTD, the downhole environment is characterized by frequent pressure fluctuations, severe mechanical vibrations, and cyclic thermal loads, all of which impose stress on sensitive electronic and mechanical assemblies. These conditions heighten the risk of tool malfunctions that may disrupt resistivity measurements, undermining the effectiveness of geosteering decisions.

Rykov (2024) provides an overview of how mechanical vibrations, particularly those associated with stick–slip, bit bounce, and lateral whirl, contribute to failures in BHA components, including resistivity sensors. The article notes that such dynamic loads can accelerate fatigue in tool housings and internal connectors, leading to microstructural damage that manifests as intermittent sensor outputs or complete electronic failure. These reliability issues often require corrective trips to replace or repair downhole tools, resulting in increased NPT and higher operational costs.

Moreover, the unpredictable nature of vibration-induced stresses complicates the ability of drilling crews to anticipate tool degradation. Subtle declines in data consistency or sudden loss of signal integrity may only become apparent after significant formation has been drilled, by which point the cost of remedial action is higher. This shows the importance of deploying robust diagnostic frameworks that can continuously monitor tool performance and flag emerging issues before they escalate into total measurement loss or force premature tool retrieval. As the demands on CTD systems expand into more challenging well profiles and geological settings, enhancing the resilience of resistivity LWD tools remains a priority for maintaining drilling efficiency.

5.1.3 Prior Approaches to Anomaly Detection in Drilling Measurements

Anomaly detection in LWD measurements plays a critical role in keeping well placement objectives, preserving tool reliability, and minimizing operational disruptions. Resistivity LWD tools provide real-time insights into formation properties that are key for geosteering decisions. Unexpected deviations in resistivity responses, phase shift readings, or attenuation patterns can indicate not only formation changes but also emerging tool malfunctions. Traditional approaches for monitoring these downhole measurements typically rely on rule-based systems or fixed thresholds. In these frameworks, values that exceed predefined operating envelopes are flagged as potential concerns. While straightforward to implement and often embedded within standard drilling monitoring platforms, these methods are limited in distinguishing real formation shifts from tool faults. They lack the capacity to recognize more complex temporal signatures or subtle precursors to sensor degradation, and they do not adapt to evolving downhole conditions such as progressive tool wear or variations in formation conductivity.

To overcome these limitations, ML techniques have increasingly been explored for detecting anomalies in well log and LWD data streams. Du et al. (2024) provide a broad overview of data-driven monitoring across geological drilling processes, noting how learning-based models can capture intricate relationships within multivariate time series that static rules cannot. Altindal et al. (2024) extended this by applying advanced anomaly detection frameworks to multivariate drilling datasets, demonstrating improved sensitivity to patterns that precede more overt failures. These approaches show the benefit of moving beyond simple thresholds toward models that learn normal measurement distributions and flag departures based on learned expectations.

Generative models have also gained attention for their ability to characterize typical measurement patterns in scenarios where labeled failure data is sparse. Al-Fakih et al. (2024) employed ensemble generative adversarial networks to enhance anomaly detection in well log data, modeling the distribution of healthy logs and identifying departures that could signal emerging tool or formation anomalies. Similarly, Koeshidayatullah et al. (2024) leveraged large-scale time-series foundation models for subsurface log prediction and anomaly detection, capturing dependencies that simpler architectures might miss. Mikołajewska et al. (2025) highlighted the use of generative AI within digital twin frameworks for predictive maintenance, which, although situated in broader industrial contexts, illustrates concepts that directly translate to downhole logging environments.

Shah et al. (2024) addressed a related challenge by using synthetic data generation to augment training sets for predicting the remaining useful life of downhole drilling sensors. Their approach is relevant to LWD diagnostics, where failure examples are often limited. By simulating degradation scenarios, they demonstrated how predictive models could better anticipate sensor decline, potentially reducing the risk of sudden tool loss or unplanned trips.

Despite these advances, much of the existing literature continues to emphasize either general well log anomalies or surface and mechanical dysfunctions, with comparatively limited focus on the health of LWD resistivity sensors used in steering operations. The principles established by these studies, such as modeling normal behavior through generative frameworks and employing reconstruction-based diagnostics, provide a strong foundation for extending anomaly detection to resistivity tool monitoring. This creates an opportunity to develop systems that can detect subtle deviations in phase shift or attenuation trends before they escalate into complete measurement failures, supporting more reliable geosteering.

5.2 Dataset Description and Preparation

5.2.1 Overview of CTD Wells and Resistivity Tool Measurements

This study utilizes a field dataset derived from four real CTD wells completed in underbalanced reservoirs. These wells were selected to capture a wide range of operational scenarios and downhole environments typical of CTD applications, including failure labels. Collectively, the dataset encompasses over three million seconds of drilling and logging data, providing a basis for developing and validating anomaly detection models.

The primary focus is on resistivity measurements acquired by LWD wave propagation tools deployed during these CTD operations. These tools recorded dual-frequency electromagnetic responses, specifically attenuation and phase shift curves, which serve as key proxies for formation resistivity. By continuously logging these measurements throughout drilling, the dataset captures both the evolving resistivity signatures of the formation and the dynamic responses of the tools to varying mechanical loads, vibration, and downhole conditions.

Notably, the dataset includes explicit interval-level annotations identifying periods of known resistivity tool malfunctions. These labeled intervals were compiled through a combination of surface diagnostics, post-run tool inspections, and cross-validation against expected formation trends, ensuring that periods of degraded measurement quality are clearly distinguished. This labeled structure supports supervised evaluation of anomaly detection performance.

Across the four wells, the data reflects a diversity of CTD operational strategies, offering a foundation for training and testing models intended to generalize across CTD campaigns. This diversity is valuable for evaluating the robustness of variational autoencoder-based frameworks under different well geometries, lithological variations, and operational loads, all of which can influence resistivity tool behavior.

5.2.2 Key Parameters: Dual-Frequency Attenuation and Phase Shift

The dataset analyzed in this study centers on four primary resistivity measurement channels acquired by a wave propagation LWD tool. Specifically, it includes dual-frequency phase shift and attenuation signals, with both low-frequency and high-frequency components recorded continuously throughout the drilling process. These measurements provide distinct but complementary perspectives on formation resistivity. The lower-frequency channel, typically around 400 kHz, achieves deeper penetration into the formation, while the higher-frequency channel, typically around 2 MHz, is more sensitive to the near-borehole zone. This dual-frequency approach enables improved characterization of formation properties by comparing responses at different investigation depths.

The operating principle of the tool is illustrated in Fig. 5.3, adapted from Macune et al. (2006). The instrument employs a four-coil layout with two transmitters placed at either end of the tool and two receivers positioned in the center. Each transmitter generates an alternating magnetic field, which induces currents in the surrounding formation. These currents in turn

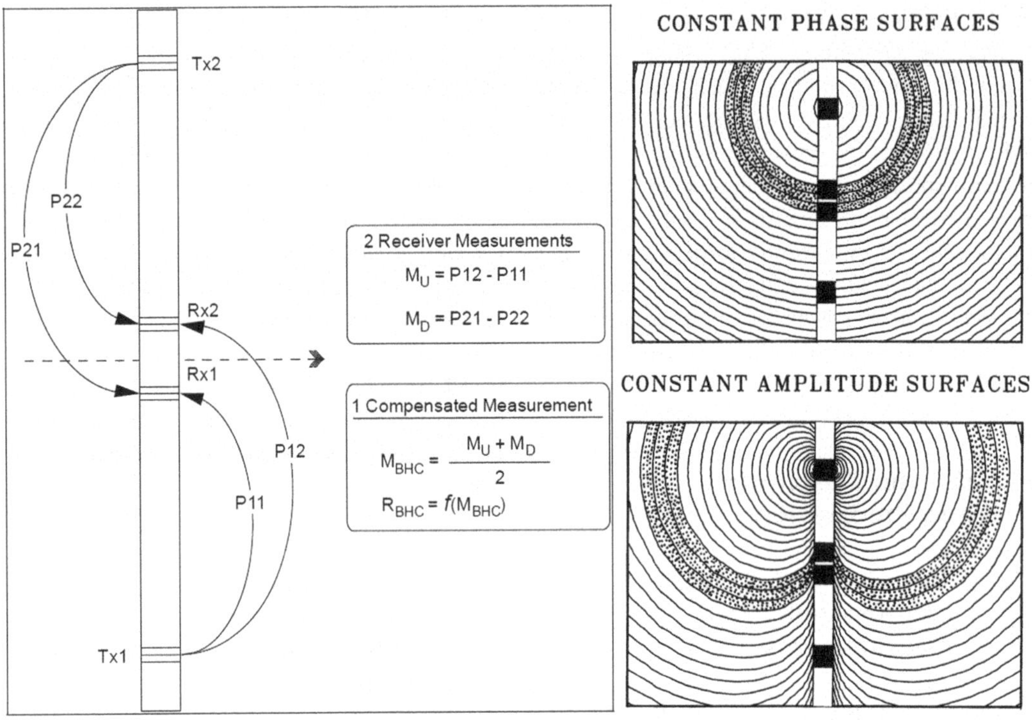

Fig. 5.3 LWD wave-propagation tool design with dual transmitters and receivers, showing phase and amplitude investigation depths (Macune et al. 2006)

create secondary electromagnetic fields that are detected by the receivers. The phase shift represents the change in phase angle between the transmitted and received electromagnetic waves as they propagate through the formation, while attenuation measures the reduction in signal amplitude caused by energy loss in the medium.

Because of the electromagnetic field behavior and the influence of skin effect and tool geometry, phase shift and attenuation measurements exhibit different depths of investigation. As shown in Fig. 5.3, the phase shift measurement is more concentrated near the borehole, capturing resistivity variations primarily in the near-wellbore zone. In contrast, the attenuation measurement responds to a broader volume, integrating resistivity effects from deeper into the formation. This difference is supported by theoretical and modeling studies, such as those by Habashy and Anderson (1991), who demonstrated that the phase shift response decays more rapidly with distance from the tool while the attenuation response remains sensitive farther into the formation.

To further mitigate borehole influences, the tool averages compensated measurements derived from the two transmitter–receiver pairs arranged in a mirror image configuration. Each pair is similarly affected by borehole irregularities but with opposing biases, so combining them reduces distortion. A dielectric correction is then applied to produce a compensated resistivity value that reflects true formation properties at the midpoint between the receivers.

The dataset used in this research focuses exclusively on these four resistivity channels. This design isolates the diagnostic challenge to the resistivity measurements themselves, emphasizing the importance of modeling the normal electromagnetic response under varying formation conditions. It also sharpens the objective of detecting deviations that signify potential sensor degradation or malfunction versus normal resistivity variations due to operational conditions.

5.2.3 Data Normalization and Preprocessing Considerations

Effective use of machine learning models, particularly unsupervised architectures like a variational autoencoder (VAE), requires careful preprocessing to support stable training and meaningful anomaly scoring. In this study, preprocessing was tailored specifically to the four resistivity measurement channels available in the dataset: dual-frequency attenuation and phase shift signals recorded by the LWD tool.

Each channel was individually scaled using a min–max normalization approach to map values into a common range. This scaling helped balance the contributions of the different resistivity channels during training and ensured that the reconstruction error used for anomaly scoring was not dominated by any single measurement type due to differences in numerical magnitude. Statistics for normalization were computed exclusively from the healthy training intervals to prevent bias from known malfunction periods.

The dataset preparation did not involve gap filling or interpolation, since intervals with discontinuities, telemetry dropouts, or known mechanical events were excluded from the training set. This decision avoided introducing artificial patterns that could confuse the unsupervised learning process. By focusing exclusively on continuous, healthy logging sequences for training, the preprocessing workflow established a clean foundation for learning normal tool behavior, which was important for achieving anomaly detection during testing on held-out malfunction intervals.

5.3 Variational Autoencoder Framework for Tool Anomaly Detection

This study developed and applied a VAE approach to detect malfunctions in LWD resistivity tools deployed during CTD operations. The objective was to establish a robust, data-driven method capable of distinguishing between healthy tool responses and sensor faults. This section details the architecture of the model, the strategy for training and anomaly scoring, the augmentation procedures used to strengthen generalization, and the cross-validation approach adopted to evaluate model performance.

5.3.1 VAE Architecture Design and Training Strategy

The anomaly detection framework developed in this study is based on a feedforward VAE configured to model pointwise resistivity tool behavior. At each drilling time step, the VAE receives a four-dimensional input vector composed of the scaled attenuation and phase shift measurements recorded at both 400 kHz and 2 MHz. These inputs were independently normalized to a [0, 1] range using min–max scaling parameters derived solely from healthy training data within each cross-validation fold. This ensured that the scaling process avoided contamination by labeled malfunction intervals and aligned with the output range of the decoder's sigmoid activation function.

The encoder network is a fully connected feedforward architecture that maps the four-dimensional input into a ten-dimensional latent space. This latent space is parameterized by vectors of means and standard deviations, from which latent samples are drawn by introducing Gaussian noise. Sampling in this latent space allows the model to capture the inherent variability of normal tool responses while encouraging a structured encoding through regular ization toward a standard normal distribution. This structure was enforced by incorporating the standard evidence lower bound objective, with the reconstruction term implemented as a mean absolute error (MAE) loss between the input and reconstructed output.

The decoder is similarly composed of feedforward layers that map the ten-dimensional latent samples back to the original four-dimensional sensor space. It uses sigmoid activations in the final layer to produce outputs consistent with the normalized input scale. The overall architecture avoids recurrent or convolutional components, reflecting the study's design focus on detecting pointwise anomalies that do not necessarily require sequential context to manifest. This simplification was supported by observations in the dataset that many resistivity tool failures produce sharp, standalone deviations in one or more channels, which are directly detectable without temporal aggregation.

Model training was conducted using the Adam optimizer with a learning rate selected through preliminary experiments to ensure stable convergence. Early stopping was employed based on reconstruction loss on a held-out portion of the training set, which prevented overfitting to features of individual wells. Each model was trained exclusively on intervals of confirmed healthy tool operation, which forced the VAE to learn a compact representation of normal resistivity tool responses.

The entire workflow was implemented in Dataiku, providing reproducibility across cross-validation folds. Figure 5.4 illustrates the schematic flow of the architecture, highlighting the encoding of normalized sensor inputs into latent variables, stochastic sampling from this latent distribution, and reconstruction back to the sensor space. This schematic is from Kingma and Welling (2019)and represents the fundamental operational structure used in this study.

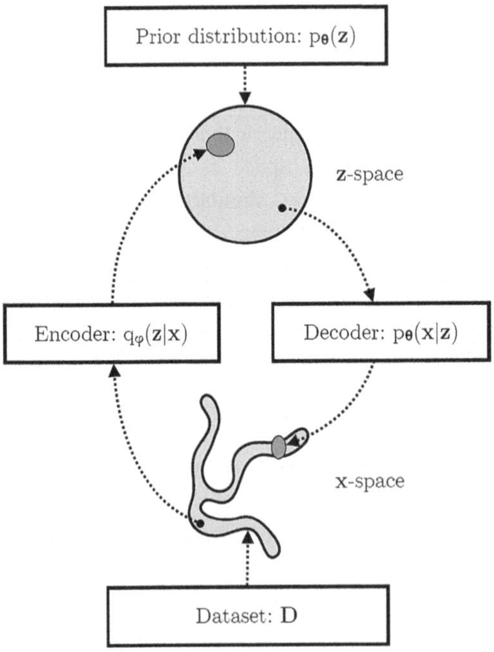

Fig. 5.4 VAE architecture showing encoding to latent space and reconstruction (Kingma and Welling 2019)

By explicitly designing the architecture around pointwise detection, with a relatively small latent dimension of ten, the model achieved a balance between capturing meaningful interrelationships among the four resistivity channels and maintaining sufficient regularization to generalize across varying well conditions. This architectural choice also ensured that the VAE remained computationally lightweight, supporting rapid inference suitable for real-time or near-real-time deployment in future applications.

5.3.2 Use of Reconstruction Error for Anomaly Scoring

The detection of resistivity tool anomalies in this framework relied on computing the reconstruction error produced by the VAE. At each time step, the model processed a four-dimensional input vector containing the scaled attenuation and phase shift measurements at 400 kHz and 2 MHz. The decoder then generated a reconstruction of this input, and the absolute differences were aggregated to quantify how well the input aligned with the learned distribution of healthy tool behavior.

If X_t denotes the input vector at time step t and $\widehat{X_t}$ its reconstruction, the reconstruction error E_t was calculated as the MAE across all four features:

$$E_t = \frac{1}{4} \sum_{i=1}^{4} \left| X_{t,i} - \widehat{X_{t,i}} \right| \tag{5.1}$$

This formulation ensured that deviations in any of the resistivity channels contributed equally to the anomaly score. MAE was selected over squared error due to its reduced sensitivity to occasional large deviations, supporting a more balanced sensitivity across diverse operating ranges.

The resulting scalar sequence E_t was then aligned with the well trajectory, allowing anomaly profiles to be visualized alongside depth or time to aid diagnostic interpretation. Elevated values of E_t indicated that the input data did not conform to the statistical patterns learned from healthy tool operation, flagging these points as potential anomalies.

To convert reconstruction errors into binary anomaly labels, a percentile-based threshold was applied. Rather than fixing this percentile manually, it was treated as a hyperparameter and optimized using Bayesian optimization within each cross-validation fold. This process systematically explored threshold settings to identify values that balanced detection sensitivity against false positive rates, adapting to the reconstruction error distribution specific to each training scenario. By doing so, the framework maintained consistency in detection criteria even when the statistical properties of the healthy data varied between wells, supporting more reliable generalization.

This approach allowed the system to identify subtle tool degradations reflected in shifts across any combination of the resistivity channels, without requiring explicit knowledge of downhole conditions or formation properties. It also provided a quantitative basis for comparing model outputs across different wells and validation folds, which was key for evaluating the robustness of the detection strategy.

5.3.3 Generative Augmentation and Its Role in Model Generalization

The VAE architecture used in this study offered a mechanism for generating synthetic data through its learned latent space. This property was employed to perform data augmentation, with the objective of enhancing the generalization of the model beyond the specific operational scenarios represented in the original well intervals.

After training the initial VAE on healthy tool data, latent vectors were sampled directly from a standard normal distribution that aligned with the prior imposed during training. Each sampled latent vector was decoded to produce a synthetic four-dimensional vector corresponding to the resistivity measurements. These synthetic samples mirrored the statistical relationships captured by the VAE during training, introducing controlled variability that remained consistent with the distribution of typical tool behavior.

This procedure allowed the effective size of the healthy training set to be expanded without requiring additional field data. Each generated sample was subjected to the same min–max normalization constraints derived from the healthy data to ensure compatibility with the scale of the decoder outputs. By blending these synthetic samples with the original data, the training distribution incorporated a broader representation of plausible healthy tool responses.

The level of augmentation, defined by the ratio of synthetic to original samples, was incorporated into the modeling pipeline as a tunable hyperparameter. Bayesian optimization was used to explore different augmentation levels systematically, determining how much synthetic expansion could be introduced while still maintaining a distribution that faithfully represented normal tool operation. This approach allowed the model training process to adaptively adjust the diversity of the healthy baseline it learned from, grounded in a probabilistic exploration rather than fixed assumptions.

All augmented datasets underwent complete retraining of the VAE under identical architecture and optimization settings. The methodology ensured that the introduction of synthetic data did not alter the fundamental structure of the model, maintaining consistency in how inputs were processed and how reconstruction errors were computed. This integration of generative augmentation into the training protocol provided a controlled means to mitigate potential overfitting to narrow patterns present in the original limited set of wells, supporting a more robust characterization of healthy resistivity tool behavior across varying downhole conditions.

5.3.4 Leave-One-Well-Out Cross-Validation Strategy

To evaluate the anomaly detection framework under realistic deployment scenarios and ensure generalization beyond the wells seen during training, a leave-one-well-out cross-validation strategy was implemented. In this approach, the dataset was partitioned such that in each fold, three of the four wells were used exclusively for training while the remaining well served as an independent test set. This process was rotated so that each well functioned as the held-out test set exactly once.

This validation design maintained strict separation between the wells used to establish the model's representation of healthy tool behavior and those used to assess anomaly detection capability. By ensuring that the VAE was always tested on data from a well it had not seen during training, the method provided a direct assessment of how effectively the learned reconstruction patterns generalized to new downhole environments, formation characteristics, and operational nuances.

Within each fold, only intervals confirmed to be free of tool malfunctions, based on engineering reviews and field reports, were included in the training set. Any intervals labeled as containing resistivity tool faults were withheld entirely and used solely for evaluation. This structure preserved the unsupervised nature of the training process, allowing the VAE to learn normal tool behavior without exposure to examples of malfunction, while still providing a mechanism to test its anomaly scoring on known fault intervals.

All hyperparameters, including the level of generative augmentation and the percentile used to set the anomaly detection threshold, were optimized independently within each training fold using Bayesian optimization. This avoided leaking information from the test well and ensured that each fold's threshold and training diversity were tuned specifically to the statistical properties of the training data drawn from the other three wells.

The reconstruction error profiles generated for the held-out well were then compared against the threshold derived in that fold to identify points flagged as anomalous. This fold-wise approach allowed a thorough assessment of the system's ability to detect tool faults under completely unseen well conditions, aligning the evaluation closely with practical scenarios where models would need to operate on wells outside the original training set. The methodology ensured that conclusions drawn about detection performance reflected true generalization across different well operations rather than artifacts of repeated exposure to the same operational contexts.

5.4 Experimental Results and Analysis

This section presents experimental findings from applying the VAE framework to the resistivity tool dataset, focusing on how the model performed under different training scenarios and how it responded to various operational conditions.

5.4.1 Effect of Data Augmentation on the Reconstruction Error Distribution

The impact of synthetic data augmentation on the behavior of the VAE reconstruction error was investigated to understand how expanding the set of healthy training examples influences the model's sensitivity and tolerance to variation. In this study, the VAE was trained using leave-one-well-out cross-validation, where healthy data from three wells were used for training and the remaining well reserved for testing. This rotation ensured that the model was always evaluated on completely unseen data.

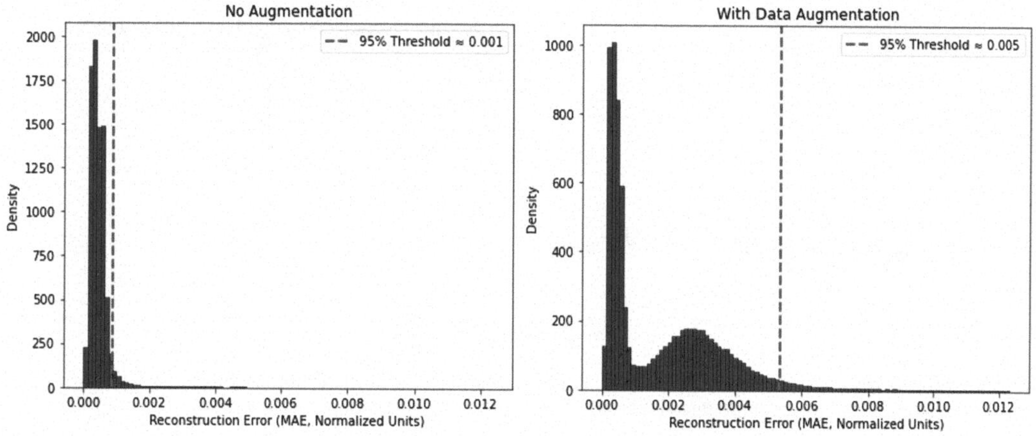

Fig. 5.5 Histograms of VAE reconstruction error (MAE) across all training wells without (left) and with (right) data augmentation. The red dashed line shows the 95th percentile threshold in normalized units (Urdaneta et al. 2025)

Figure 5.5 illustrates how the inclusion of synthetic samples alters the distribution of reconstruction errors across the training data. Both panels show histograms of the MAE computed between the four-channel input vectors and their reconstructions, expressed in normalized units following MinMax scaling of each channel to the 0 to 1 range. This normalization was performed separately for each cross-validation fold using only the healthy training data, ensuring that the scale of the reconstruction error remained consistent and unbiased by test data.

The left panel of Fig. 5.5 shows the error distribution for the baseline model trained solely on real data. The histogram exhibits a narrow, sharply peaked profile with most reconstruction errors tightly clustered near zero. This indicates that the VAE learned to represent the limited range of healthy behaviors present in the real training set with high fidelity, leaving little tolerance for even modest deviations. The 95th percentile threshold, computed using Bayesian optimization to balance detection sensitivity and false positive rates, is approximately 0.001 in normalized units. This low threshold reflects the constrained variability learned from purely real examples and means that minor departures from the narrow learned manifold might trigger anomaly flags.

In contrast, the right panel of Fig. 5.5 shows the reconstruction error distribution after introducing synthetic samples generated by decoding latent vectors sampled from a standard normal distribution. These synthetic inputs, treated as additional healthy examples, expanded the diversity of normal patterns encountered during training. As a result, the reconstruction error histogram becomes broader, with a noticeable secondary mode appearing at slightly higher error levels. The corresponding 95th percentile threshold shifts to about 0.005 in normalized units. This higher threshold allows the model to accept a wider range of variations as normal, reducing the likelihood of misclassifying benign fluctuations as faults.

By training with synthetic samples, the VAE learns a more comprehensive representation of acceptable tool behavior, improving robustness to operational variability. At the same time, because the underlying latent structure still isolates truly anomalous patterns, the system maintains the ability to identify genuine malfunctions. This tradeoff between sensitivity and specificity is key for field deployment, where false positives can lead to unnecessary operational interruptions, while missed detections risk propagating incorrect tool readings into geosteering decisions.

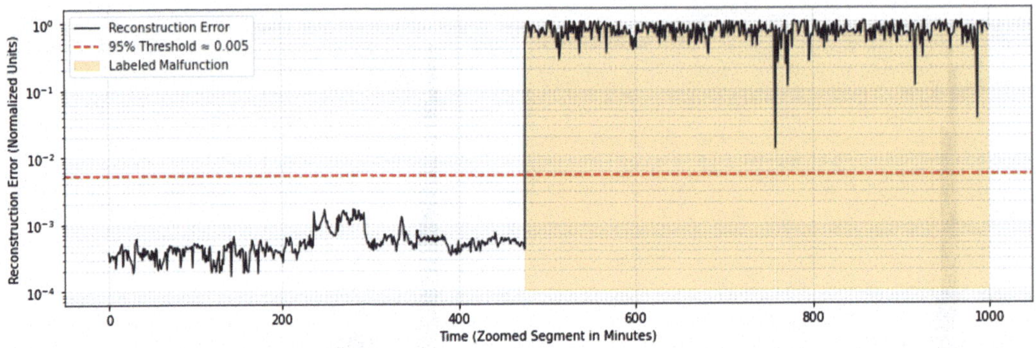

Fig. 5.6 VAE reconstruction error during a segment of well data, showing labeled malfunctions (orange) and predicted anomalies based on a 95th percentile threshold (red) in normalized units (Urdaneta et al. 2025)

5.4.2 Detection of Known Tool Malfunction Intervals

The practical value of the reconstruction error approach becomes clear when applied to real drilling sequences that include known tool malfunctions. Figure 5.6 illustrates this scenario using a representative segment of well data that includes a documented resistivity tool failure. In this plot, the black curve shows the VAE reconstruction error over time, plotted in normalized units on a semi-logarithmic scale to highlight both small and large deviations. The red dashed line marks the 95th percentile threshold established during training. The orange shaded region indicates the interval identified through post-job engineering analysis as a period of confirmed tool malfunction.

Prior to the onset of the malfunction, the reconstruction error remains relatively stable and consistently below the threshold, indicating that the tool measurements align well with the patterns the VAE learned as healthy. Once the tool enters the failure mode, the reconstruction error rises sharply and remains elevated, frequently exceeding the threshold throughout the duration of the malfunction.

This behavior demonstrates the model's ability to detect the onset and persistence of tool degradation without relying on explicit supervised labels during training. The VAE flags these intervals solely because the tool signals during the malfunction deviate substantially from the manifold of normal operation learned during training. This approach is particularly suitable for drilling operations, where labeled fault examples are rare and not always consistent, making unsupervised methods valuable for real-time anomaly detection.

5.4.3 Latent Space Analysis Using t-SNE Projections

To gain further insight into how the VAE internally organizes healthy versus anomalous tool behaviors, the latent representations learned by the encoder were examined. Each four-dimensional resistivity input sample was mapped to a ten-dimensional latent vector. These latent encodings were then projected into two dimensions using t-distributed Stochastic Neighbor Embedding (t-SNE), a technique designed to preserve local structure from high-dimensional spaces.

Figure 5.7 presents this t-SNE visualization. Blue points correspond to input samples drawn from healthy segments of the wells used during training, while red triangles represent samples from known malfunction intervals that were excluded from the training process. The clear spatial separation between normal and faulty inputs suggests that the VAE has developed an internal latent representation that meaningfully distinguishes typical tool dynamics from abnormal patterns associated with malfunctions.

Unlike simple linear techniques like principal component analysis, t-SNE focuses on main-

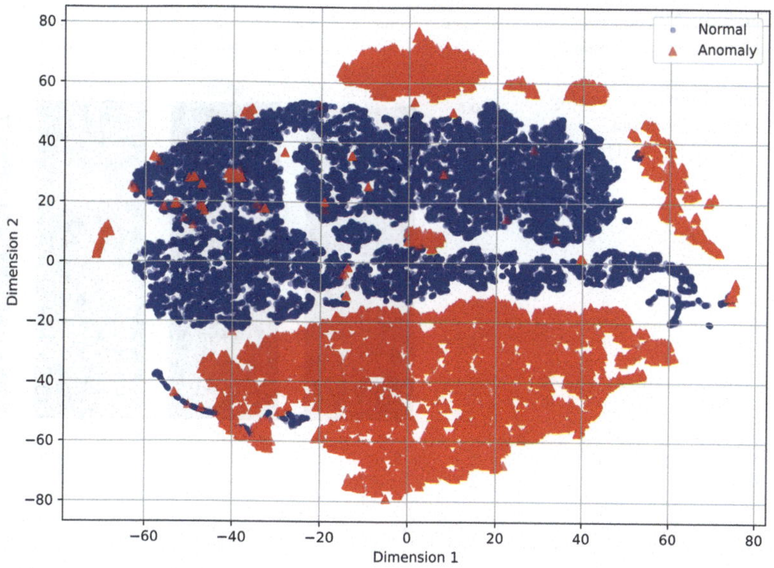

Fig. 5.7 t-SNE projection of the VAE latent space. Blue points represent normal input encodings from training wells; red triangles show encodings of test-time anomalies (Urdaneta et al. 2025)

taining neighborhood relationships, meaning points that are close in the original ten-dimensional latent space remain close in the two-dimensional embedding. The resulting clusters illustrate that even though the model was never explicitly trained to classify faults, it naturally encodes them in regions of latent space distinct from normal operation. This further supports the use of reconstruction error as a diagnostic indicator, since the anomalous signals are fundamentally different in how they are internally represented.

5.4.4 Performance Across Wells Under Varying Augmentation Levels

The influence of synthetic data augmentation on the overall performance of the anomaly detection framework was evaluated using leave-one-well-out cross-validation. This strategy tested the model's ability to identify resistivity tool malfunctions in a completely held-out well after training exclusively on healthy data from the remaining three wells. By rotating the held-out well in each fold, a comprehensive assessment was achieved across all four CTD wells in the dataset.

Performance was quantified using the F1 score, which provides a balanced metric that incorporates both precision, the proportion of flagged anomalies that were true tool faults, and recall, the proportion of actual tool malfunctions correctly identified by the model. This metric is particularly appropriate in scenarios such as drilling tool health monitoring, where both missed detections and excessive false positives carry operational costs.

The analysis explored augmentation levels ranging from the baseline (1×, with no synthetic data) up to a 16× expansion of the original healthy dataset through generated samples. For each level of augmentation, the training dataset was augmented by decoding latent vectors sampled from the standard normal distribution, producing synthetic inputs consistent with the VAE's learned representation of healthy tool signals.

Results of this cross-validated performance study are summarized in Fig. 5.8, which shows the average F1 scores achieved across the wells under different augmentation levels. Without any augmentation, the baseline model already performed well, reaching an average F1 score of approximately 0.802. This demonstrated that even with only the original healthy data, the VAE could effectively differentiate the clear

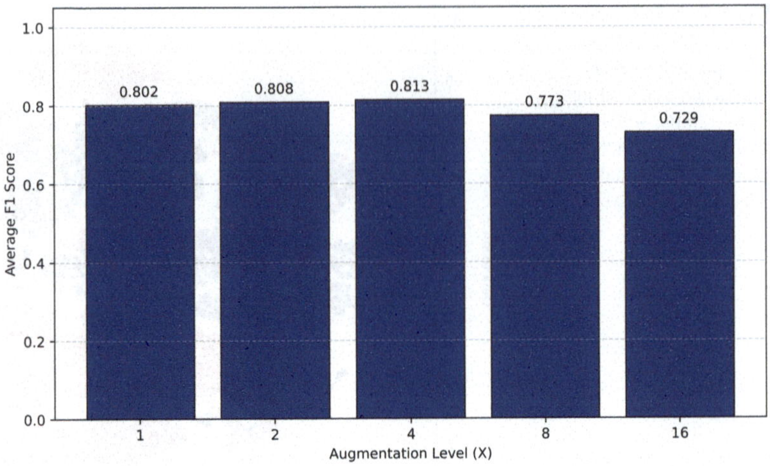

Fig. 5.8 Average F1 score across wells under varying augmentation levels (1× to 16×), using leave-one-well-out cross-validation (Urdaneta et al. 2025)

deviations associated with known resistivity tool failures.

Introducing moderate levels of augmentation led to consistent improvements. At 2× augmentation, the average F1 score rose to 0.808, while at 4× augmentation it peaked at 0.813. This pattern suggests that expanding the training set with additional generated examples of healthy behavior improved the model's generalization to previously unseen yet benign variations across different wells. By learning a broader manifold of what constitutes normal tool responses, the VAE became less prone to incorrectly flagging harmless fluctuations that simply reflected variability in drilling conditions.

However, when the augmentation level was increased further, performance began to decline. At 8× augmentation, the average F1 score dropped to 0.773, and at 16× it decreased further to 0.729. This trend indicates that excessive augmentation can dilute the precision of the learned boundary between healthy and faulty states. By flooding the training distribution with large volumes of synthetic data, the model may overexpand its definition of normal tool behavior, becoming too tolerant and thereby less sensitive to subtle signatures of incipient tool degradation.

These findings emphasize the importance of carefully tuning the degree of augmentation. While synthetic examples generated by the VAE are structurally consistent with the latent distribution learned from actual tool data, excessively large volumes of generated samples risk overrepresenting artificial patterns that might not correspond to the nuanced realities of downhole operations. This can weaken the model's discriminative capacity by flattening distinctions between genuine malfunctions and complex but normal signal variations.

5.5 Conclusion

This chapter investigated the use of a VAE framework for unsupervised detection of resistivity tool malfunctions during CTD operations. The approach relied solely on healthy intervals for training, allowing the model to learn patterns of normal sensor behavior without exposure to explicit failure examples. This unsupervised setup is particularly valuable for downhole applications, where detailed labeled fault datasets are rarely available.

The core methodology involved training the VAE on multichannel resistivity data collected from four CTD wells. Each input sample was a one-second segment capturing simultaneous readings from the tool's four resistivity-related channels. By mapping these four-dimensional inputs into a ten-dimensional latent representation and reconstructing them back, the VAE learned a transformed feature space in which deviations could be quantified through reconstruction errors. A leave-one-well-out cross-validation strategy was employed to evaluate generalization, ensuring that the model was always tested on an entirely unseen well. This provided a realistic measure of the model's robustness across varying downhole conditions.

To support detection, a dynamic threshold was established at the ninety-fifth percentile of reconstruction errors computed on the training data. This threshold was recalculated for each cross-validation fold and level of data augmentation, ensuring that it adapted to the distribution of healthy reconstruction errors. Bayesian optimization was employed to jointly tune the latent space dimension and the percentile threshold, with the objective of balancing sensitivity to true tool malfunctions against robustness to ordinary signal variability. This process identified a latent space of ten dimensions as an effective trade-off that maintained computational efficiency while capturing meaningful relationships among the resistivity channels. It also confirmed that using the ninety-fifth percentile as a detection threshold provided an appropriate compromise between minimizing false positives and maintaining reliable detection of emerging tool faults for this dataset.

The incorporation of generative data augmentation further enhanced the model's capabilities. By sampling latent vectors from a standard normal distribution and decoding them into synthetic inputs, the training set was expanded in a controlled manner. This strategy exposed the VAE to a broader spectrum of normal operational variations, which proved beneficial up to moderate levels of augmentation. The resulting broader reconstruction error distribution raised the detection threshold, improving the model's tolerance to benign fluctuations without sacrificing its ability to flag genuine anomalies.

As shown in Fig. 5.5, comparing reconstruction error distributions with and without augmentation revealed that moderate augmentation widened the error spread, increasing the threshold to better encompass natural operational variability. This reduced the likelihood of false positives under slightly altered drilling or formation conditions. Figure 5.6 then demonstrated how this translated into time-aligned detection, where the model maintained low reconstruction errors during healthy intervals and exhibited a marked rise precisely aligned with known malfunction periods. This illustrates the practical effectiveness of the method for identifying tool issues without prior exposure to fault data.

In addition, analysis of the learned latent space using t-SNE projections, shown in Fig. 5.7, revealed clear clustering of normal behavior apart from anomaly intervals. This provided an interpretable view of how the VAE organized its internal representations, reinforcing that it not only reconstructed inputs but also captured meaningful distinctions between healthy and malfunctioning tool responses.

Quantitative performance trends under varying levels of augmentation, presented in Fig. 5.8, further supported these conclusions. The model achieved an average F1 score of 0.802 without augmentation, which modestly improved to a peak of 0.813 at a fourfold augmentation level. Beyond this point, performance declined, likely because excessive synthetic data expanded the model's conception of normal behavior too broadly, reducing its sensitivity to real faults. This outcome shows the importance of controlled augmentation and careful tuning to strike the right balance between generalization and fault sensitivity.

In practical terms, this methodology offers a lightweight yet effective solution for early detection of resistivity tool malfunctions. By operating on one-second multichannel inputs, the framework is well suited for real-time deployment, providing timely alerts that can prevent misinformed geosteering decisions. This contributes to maintaining wellbore placement within target zones, ultimately supporting improved production outcomes.

References

Al-Fakih, A., Koeshidayatullah, A., Mukerji, T., & Kaka, S. I. (2024). Enhanced anomaly detection in well log data through the application of ensemble GANs. arXiv preprint arXiv:2411.19875.

Altindal, M. C., Nivlet, P., Tabib, M., Rasheed, A., Kristiansen, T. G., & Khosravanian, R. (2024). Anomaly detection in multivariate time series of drilling data. Geoenergy Science and Engineering, 237, 212778.

Du, S., Huang, C., Ma, X., & Fan, H. (2024). A review of data-driven intelligent monitoring for geological drilling processes. Processes, 12(11), 2478.

Habashy, T., & Anderson, B. I. (1991). Reconciling differences in depth of investigation between 2-MHz phase shift and attenuation resistivity measurements. In SPWLA Annual Logging Symposium. SPWLA.

Hermawan Manuaba, I. B. G., Aljishi, M., Van Steene, M., & Dolan, J. (2024). Logging-while-drilling laterolog vs. electromagnetic propagation measurements: Which is telling the true resistivity? SPE Journal, 29(8), 4000–4013.

Kingma, D. P., & Welling, M. (2019). An introduction to variational autoencoders. Foundations and Trends in Machine Learning, 12(4), 307–392.

Koeshidayatullah, A., Al-Fakih, A., & Kaka, S. I. (2024). Leveraging time-series foundation model for subsurface well logs prediction and anomaly detection. arXiv preprint arXiv:2412.05681.

Macune, D. T., Flanagan, W. D., Choi, E., & Marcellus, E. (2006). A compact compensated resistivity tool for logging while drilling. In SPE/IADC Drilling Conference and Exhibition. SPE.

Mikołajewska, E., Mikołajewski, D., Mikołajczyk, T., & Paczkowski, T. (2025). Generative AI in AI-based digital twins for fault diagnosis for predictive maintenance in Industry 4.0/5.0. Applied Sciences, 15(6), 3166.

Rykov, Y. (2024). Drilling and logging equipment reliability in a downhole vibration environment. Journal of Petroleum Technology.

Shah, A. B., Wen, Y., Chen, J., Wu, X., & Fu, X. (2024). A deep learning approach to predicting remaining useful life for downhole drilling sensors using synthetic data generation. In Unconventional Resources Technology Conference. URTeC.

Urdaneta, C., Jeong, C., Wu, X., & Chen, J. (2025). Detecting resistivity tool malfunctions in coiled tubing drilling using generative artificial intelligence. In Unconventional Resources Technology Conference. URTeC.

6 Improving Electromagnetic Telemetry Signal-To-Noise Ratio

6.1 Background and Related Work

6.1.1 Overview of Electromagnetic Telemetry Challenges

Electromagnetic (EM) telemetry is widely employed for communication between downhole tools and surface systems due to its operational flexibility and cost benefits. Unlike acoustic telemetry, EM signals do not rely on fluid columns and therefore do not attenuate as rapidly in certain well conditions. Moreover, EM telemetry avoids the need for pressurized mud or cables, distinguishing it from mud pulse and optical systems, and enabling its deployment even under openhole conditions or during service pipe disconnections (Carpenter 2015). These advantages make EM telemetry particularly attractive for underbalanced drilling, where it is necessary to keep formation pressures higher than wellbore pressures.

However, EM telemetry also faces critical challenges. Attenuation of EM waves is strongly influenced by formation resistivity and well depth. High-conductivity layers and extended depths lead to significant energy losses, causing reduced signal-to-noise ratio (SNR) on the surface. Additionally, complex downhole assemblies containing multiple sensors and motors contribute to local EM noise. On the surface, motors, solenoids, and power systems further introduce noise into the received signals. As a result, the acquired telemetry often contains a weak downhole signal buried in substantial coherent and incoherent noise, especially within the low-frequency band of 1–25 Hz that characterizes typical EM telemetry systems.

Figure 6.1 clearly illustrates the basic EM telemetry system, showing the downhole antenna transmitting signals to surface stakes that detect voltage differentials. The figure emphasizes how the combination of attenuated signals and significant rig noise results in low SNR, which complicates reliable telemetry reception.

6.1.2 Conventional Approaches and Limitations

Traditional approaches to address low SNR in EM telemetry generally involve hardware-heavy or filtering strategies. One method is the installation of additional cabled probes in offset wells or the deployment of EM relay stations to boost signal integrity (Jannin et al. 2017; Champion 2016). While effective in certain contexts, these solutions come with considerable cost and logistical complexity.

On the signal processing side, conventional approaches rely on bandpass filtering to suppress noise outside the EM telemetry band. However, the signal and noise frequently occupy the same frequency range, which limits the utility of frequency-based filtering. Adaptive noise

C. Urdaneta et al., *Deep Learning Applications for Drilling Performance: Forecasting, Diagnostics, Telemetry, and Predictive Maintenance*, Advances in Oil and Gas Exploration & Production,
https://doi.org/10.1007/978-3-032-26648-4_6

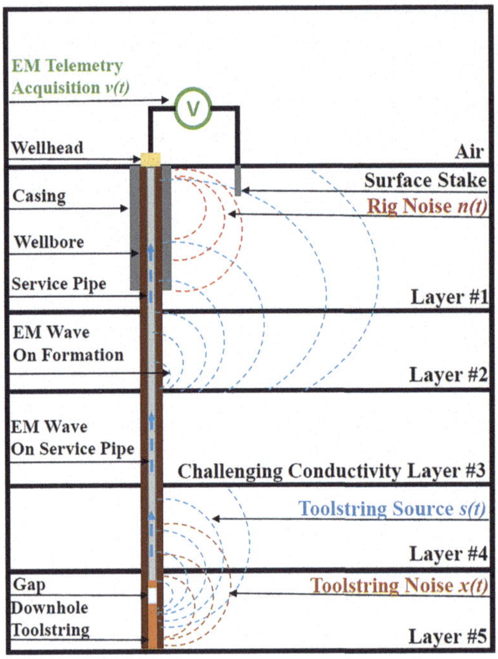

Fig. 6.1 Downhole tool EM telemetry sketch. Attenuated signals, and the large noise present, cause a low SNR in the EM telemetry (Urdaneta et al. 2023)

cancellation methods, such as those described by Keman (2016), attempt to model and subtract predictable noise patterns. Yet they often fail when noise is neither stationary nor well-characterized, as is common in dynamic drilling environments.

The inability of these traditional filters to handle overlapping frequency content is problematic. As shown in later sections of this study, both coherent and random noise frequently align with the EM telemetry frequency band, meaning classical techniques provide insufficient isolation of the desired signal.

6.1.3 Deep Learning Applications to Noise Attenuation and Source Separation

Given these limitations, DL methods have been explored as an alternative due to their ability to learn complex, data-driven representations that traditional algorithms cannot capture. Prior studies have shown that neural networks, particularly autoencoders, can perform powerful denoising and source separation tasks. Saad and Chen (2020) successfully demonstrated the application of a deep denoising autoencoder (DDAE) for seismic random noise attenuation, training models that could extract coherent signals from noisy 2D seismic inputs. Similarly, Yang et al. (2021) advanced this work by introducing skip connections and patching strategies in an unsupervised architecture to improve feature learning and generalization, addressing challenges of insufficient labeled data.

In related fields, Luo et al. (2020) developed the dual-path recurrent neural network (DPRNN), originally aimed at time-domain single-channel speech separation. This architecture decomposes long sequential data into manageable segments, processing local and global dependencies through stacked intra- and inter-chunk recurrent layers. Its demonstrated ability to disentangle overlapping temporal features has motivated exploration in broader time series applications.

The application of these DL techniques to EM telemetry aims to leverage their success in other low-SNR, multi-source contexts such as speech and seismic data. The premise is that the models can learn to represent the underlying downhole signal while attenuating or separating various types of noise, even when the noise shares significant spectral overlap with the target signal.

6.1.4 Architectures Explored: DDAE, DDUL, DPRNN

This study evaluated three architectures to enhance EM telemetry SNR: the DDAE, DDUL (deep-denoising unsupervised learning), and DPRNN, each represented in Fig. 6.2.

The DDAE was initially proposed for seismic applications by Saad and Chen (2020). It uses a two-phase process. The first is supervised, leveraging synthetic labeled datasets where the network learns to map noisy inputs to clean targets by minimizing the root mean square error (*RMSE*), defined as

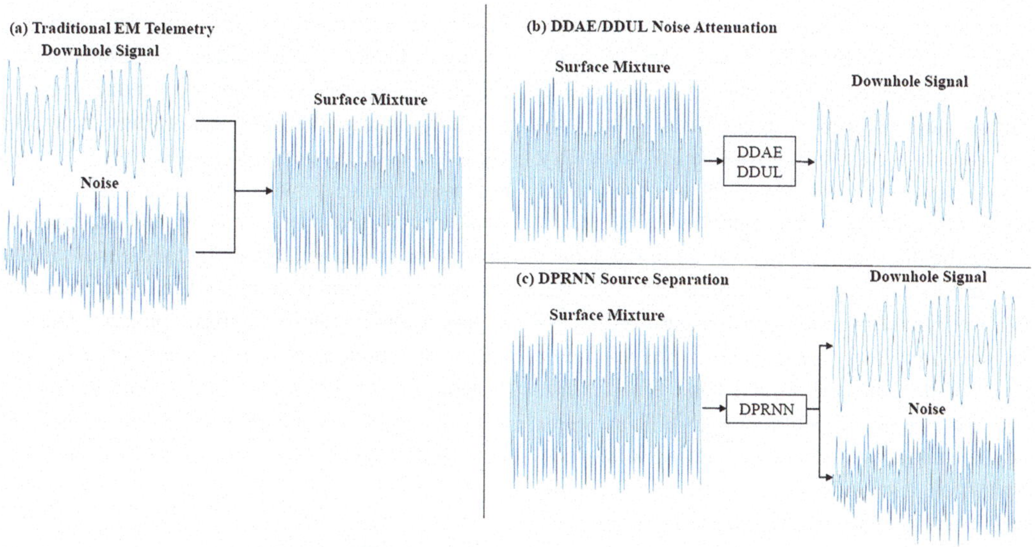

Fig. 6.2 Three DL-based architectures explored to overcome the low SNR problem. **a** The traditional EM telemetry surface signal consists of a mixture of a downhole signal and noise; **b** DDAE and DDUL noise attenuation methods filter the noise from the surface mixture; **c** DPRNN separates the signal and the noise from the input mixture (Urdaneta et al. 2023)

$$RMSE = \sqrt{\frac{1}{N}\sum_{i=1}^{N}(y_i - \hat{y}_i)^2} \quad (6.1)$$

where y_i represents the true signal value, $\hat{y}_i$ represents the model prediction, and N is the total number of samples. The second phase transitions to unsupervised training on field data, using the network to reconstruct the input itself while minimizing correlation with removed noise, thus refining its capacity to separate signal from environment-specific noise without explicit labels.

The DDUL, introduced by Yang et al. (2021), builds on the DDAE structure by adding skip blocks between encoder and decoder layers and employing overlapping patches of training data. These enhancements improve feature extraction and mitigate overparameterization risks. The DDUL operates in a fully unsupervised regime, which is advantageous for wellsite applications where clean training labels are rare.

The DPRNN from Luo et al. (2020) departs from the autoencoder structure, instead segmenting input sequences into overlapping chunks processed by intra-chunk and inter-chunk bidirectional RNNs. The overlap-add reconstruction at the output stage rebuilds the continuous signal from these processed chunks. This design allows the DPRNN to effectively handle long EM telemetry sequences by focusing on both short-term and long-term dependencies, making it promising for source separation even when signal and noise frequencies overlap.

6.2 Dataset Description and Preparation

6.2.1 Field Data Acquisition and Stake Configuration

The field dataset used in this study was collected during operational electromagnetic telemetry runs conducted by SLB. The data represent surface stake measurements acquired during routine drilling operations and are confidential. As a result, the raw dataset cannot be released publicly. However, the acquisition setup, preprocessing workflow, and evaluation methodology are described in this section to enable reproduction of the modeling approach using comparable datasets.

The experimental datasets used in this study combine real-world field measurements with synthetic signals designed to facilitate supervised training and controlled evaluations. The field data were collected during routine EM telemetry operations, leveraging standard surface stake configurations deployed around the wellhead.

Specifically, the field dataset comprised over 1.16 million data points, sampled at 160 Hz, which provided high temporal resolution necessary to characterize low-frequency EM telemetry signals within the operational band of 1–25 Hz. Data were acquired simultaneously from eight surface stakes positioned at varying distances, resistances, and orientations relative to the wellhead. This stake layout was intentionally diverse, mirroring practical installations where local terrain, accessibility, and operational constraints dictate stake placement. Figure 6.3 illustrates this setup, detailing the distance (D) and resistance (R) values associated with each stake.

These diverse stake conditions introduce variations in signal amplitude and phase as well as the noise profile recorded at each site. The multi-stake recordings were key for studying how surface array geometry impacts DL denoising and source separation performance. This aspect is especially relevant for evaluating the generalizability of learned models across installations that differ in layout and local ground conductivity.

Additionally, the field recordings naturally captured both rig-generated noise from motors, solenoids, and power lines, and downhole noise introduced by other tools and mechanical processes. Together, these factors ensured that the raw field signals contained realistic mixtures of target EM telemetry signals, coherent interference, and random noise. This provided a solid foundation for developing and testing denoising methods capable of handling complex wellsite environments.

6.2.2 Synthetic Data Construction for Supervised Learning

While the field data captured realistic noise environments, training a supervised DL model for noise attenuation necessitates a clean target, which is typically unavailable in actual telemetry logs. To address this, the study constructed a synthetic dataset by mixing real field recordings with artificially generated EM telemetry signals.

#	D (m)	R(Ω)
1	55	28.6
2	87	51.8
3	60	423
4	87	25.3
5	86	8.3
6	112	23
7	32	11.3
8	80	18

Fig. 6.3 Distance (D), resistance (R), and position map of the stakes from the input data set. Each stake had a different distance from the wellhead and a different resistance due to wellsite features (Urdaneta et al. 2023)

The synthetic EM telemetry source signals were designed to closely resemble operational downhole telemetry, using quadrature phase shift keying (QPSK) modulation to encode data with carrier frequencies between 8 and 20 Hz and bit rates around 8 bits per second. Each synthetic waveform had a controlled amplitude, typically set at 1 mV, enabling precise experimentation on how signal strength influences model performance.

The generation process is formalized in Equation 6.2:

$$s(t) = \sqrt{2E_s f_s} cos\left(2\pi f_c t + (2k-1)\frac{\pi}{4}\right),\quad k = 1, 2, 3, 4 \tag{6.2}$$

where E_s represents the energy per symbol, f_s the symbol rate, and f_c the carrier frequency. This equation ensured that the synthetic source captured the phase and frequency characteristics typical of telemetry operations.

These synthetic signals were then injected into the real field noise recordings across all eight stakes, producing composite inputs v(t) that aligned with the multi-stake acquisition geometry. This approach guaranteed that the mixture retained the spatial and spectral properties of field noise while embedding a controlled downhole signal. It also ensured that all stakes simultaneously received the synthetic signal, maintaining phase coherence and allowing the network to learn multi-stake correlations key for spatial denoising.

This hybrid dataset served two purposes: it facilitated supervised learning by providing a known clean target, and it enabled studies on the effects of frequency separation, source amplitude, and stake array geometry under controlled conditions.

6.2.3 Preprocessing and Normalization

Minimal preprocessing was applied to preserve important signal characteristics needed for accurate noise attenuation and source separation. Given that the EM telemetry band of interest lies between 1 and 25 Hz, a conventional bandpass filter was first applied to all inputs. This preprocessing step removed out-of-band energy, simplifying the subsequent learning task by eliminating irrelevant high-frequency and direct current (DC) components. It also allowed direct comparisons with traditional filter methods, establishing a baseline for evaluating DL improvements.

Beyond this initial filtering, no aggressive noise suppression or smoothing was performed. Maintaining the raw fluctuations in the data was needed to allow the neural networks to learn to distinguish signal features from noise based on the training objectives. This was particularly important for the DPRNN experiments, where temporal patterns and amplitude variations across long sequences inform the separation process.

Normalization was applied to stabilize training and ensure numerical consistency. Inputs were standardized in amplitude, to prevent large scale disparities that could bias network optimization toward certain stakes or frequencies. However, scaling factors were computed to avoid leaking information between training and test sets, preserving the integrity of cross-validation experiments.

6.3 Deep Learning Framework for Improved Telemetry Reliability

6.3.1 Deep Learning Architecture Design and Training Strategy

This study explored three principal DL architectures tailored to improve the SNR of EM telemetry data: the DDAE, the DDUL, and the DPRNN models. Each was selected based on prior successes in analogous domains such as seismic noise attenuation and speech source separation, then adapted to address the unique low-frequency, low-SNR characteristics of EM telemetry.

The DDAE, proposed by Saad and Chen (2020) for seismic denoising, combines supervised and unsupervised phases. Initially, it uses synthetic mixtures with known clean signals for supervised learning, minimizing the RMSE between the network output and the clean synthetic target. This phase initializes the network with weights that capture essential features of the telemetry signal. Following this, the DDAE transitions to an unsupervised stage where the input data is also the target. This trains the network to refine internal representations that consistently reconstruct coherent signal features while suppressing incoherent noise. Figure 6.4 illustrates this architecture, showing the 2D input matrix structure with time and stake as dimensions and the mirroring encoder-decoder layers that compress and reconstruct the data.

The DDUL, as described by Yang et al. (2021), extends the DDAE framework by introducing skip connections between encoder and decoder layers and implementing a patching technique. The skip connections facilitate the flow of fine-grained features directly across the network depth, improving reconstruction fidelity without over-reliance on deep stacked layers that increase parameter count and risk overfitting. The patching technique overlaps input data segments, expanding the dataset and promoting robustness to local variations which is a valuable adaptation given the limited quantity of clean telemetry training data typically available. Figure 6.5 provides a schematic overview of the DDUL's design enhancements over the baseline autoencoder.

The third architecture explored was the DPRNN, proposed by Luo et al. (2020) for time-domain speech separation. Unlike the DDAE and DDUL, DPRNN does not rely on an explicit autoencoder structure. Instead, it leverages a segmentation strategy that breaks long sequential inputs into overlapping chunks, processes them through bidirectional recurrent networks to learn local (intra-chunk) and global (inter-chunk) dependencies, and reconstructs the cleaned output via overlap-add synthesis. Figure 6.5 from the study details this architecture. While originally developed for speech, the DPRNN's ability to handle extended sequences made it well-suited to capturing long-range temporal relationships inherent in EM telemetry signals acquired over extended drilling intervals (Fig. 6.6).

Training strategies were aligned with the objectives of each architecture. For the DDAE, initial supervised training used synthetic

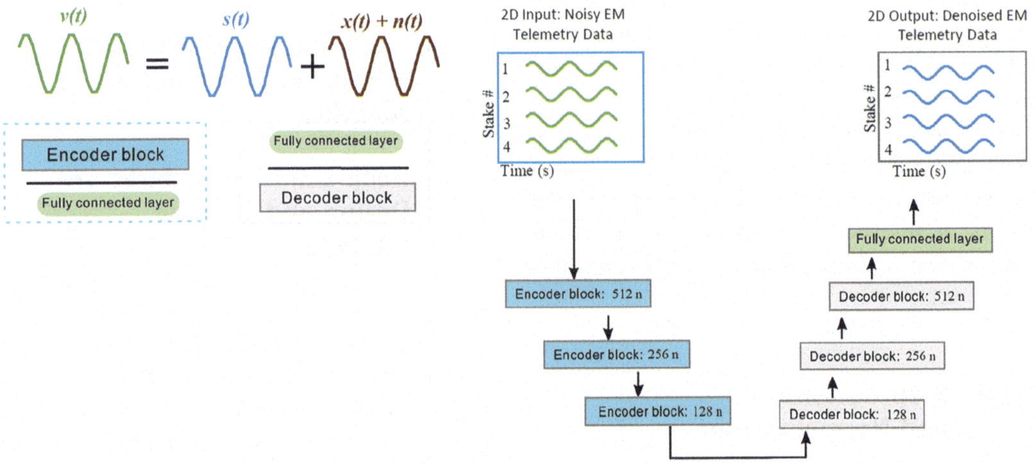

Fig. 6.4 First architecture explored, proposed by Saad and Chen (2020), for DDAE. The input data fed to the model are a 2D matrix of noisy EM telemetry data acquired at the surface with time and stake number as the two dimensions. Similarly, the output data from the model are a 2D matrix of denoised EM telemetry data with time and stake number as the two dimensions (Urdaneta et al. 2023)

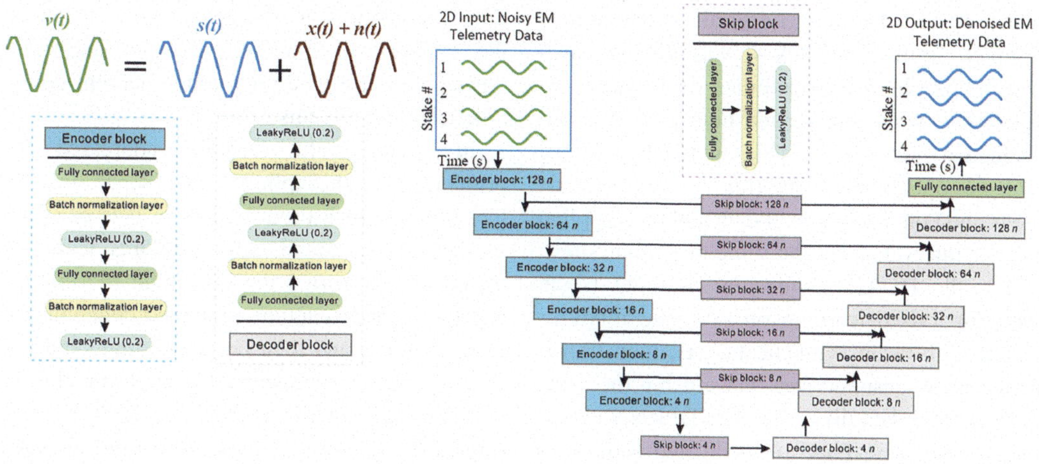

Fig. 6.5 Second architecture explored, proposed by Yang et al. (2021), for the DDUL. The authors based the DDUL on the DDAE structure, but they added skip blocks between the encoder and decoder sections of the model, and they also applied a patching technique that structures the training data to produce overlap (Urdaneta et al. 2023)

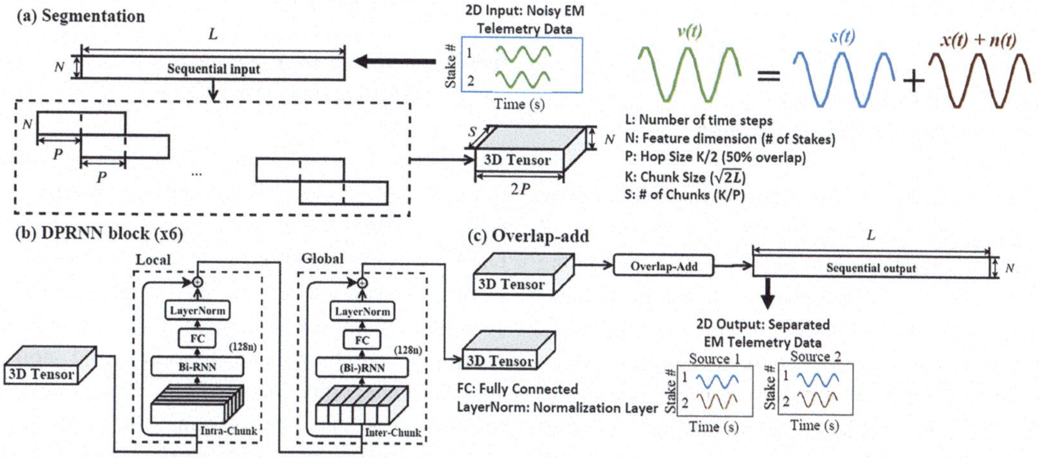

Fig. 6.6 Third architecture explored, proposed by Luo et al. (2020), for DPRNN. It consists of three stages: **a** segmentation; **b** DPRNN block; and **c** overlap-add (Urdaneta et al. 2023)

mixtures to guide the network toward recovering clean telemetry waveforms. This was important for teaching the network the core structure of QPSK-modulated signals within the typical telemetry band. The subsequent unsupervised phase allowed the network to adapt to the statistical properties of real field noise. The DDUL, by design, operated fully in an unsupervised manner, training directly on the noisy inputs to develop a representation that maximized the coherence of recovered signals. The DPRNN used supervised separation objectives, attempting to disentangle the synthetic telemetry component from the composite mixture fed into the network.

6.3.2 Data Splits and Cross-Validation Strategy

To evaluate generalization and prevent overfitting, the experiments used a clear chronological split of the field dataset. The dataset, which consisted of over 1.16 million data points sampled from eight stakes, was divided such that the first hour provided data for training and validation, while the second hour was reserved exclusively for testing. This approach reflected realistic operational conditions, where models are trained on existing well data and then applied to later intervals from the same or similar sites.

For the DDAE architecture, training began with supervised learning on synthetic mixtures paired with clean synthetic targets, then transitioned to an unsupervised phase using the same general field data divisions. The DPRNN experiments similarly relied on dedicated sequences drawn from the first hour, including uninterrupted 12-min segments, to support the model's dependence on processing long continuous sequences through its segmentation and recurrent blocks.

6.3.3 Loss Functions and Training Phases

The choice of loss functions and staged training approaches was tailored to the objectives of each architecture. For the DDAE, the training followed a two-phase process. In the first phase, a supervised loss function minimized the RMSE between the network's denoised output and a clean synthetic EM telemetry source. This use of RMSE, penalized large reconstruction deviations and helped the network learn to reproduce characteristics of the downhole signal. These supervised weights then served as initialization for the second, unsupervised phase, which minimized the correlation between the denoised output and the removed noise component. This encouraged the model to suppress incoherent noise without requiring labeled clean targets, while still leveraging prior supervised learning through transfer learning.

For the DDUL architecture, the training was conducted entirely without labeled clean references. Instead, the input surface measurements were used both as input and target, with the architecture relying on its internal feature abstraction and skip connections to minimize reconstruction discrepancies. This guided the network to filter out random and non-repeatable noise while retaining coherent downhole telemetry signals.

In contrast, the DPRNN addressed the problem through explicit source separation. Its loss function measured the RMSE between each of the two separated output streams with one aimed to reconstruct the downhole tool signal, and the other to capture combined coherent and random noise, and their corresponding synthetic ground truth components. By applying this to both outputs simultaneously, the DPRNN was trained to achieve a clean partition between desired telemetry signals and interfering noise.

6.4 Experimental Results and Analysis

6.4.1 Impact of Frequency Separation on DDAE

The influence of frequency separation between the downhole source signal and the interfering noise on the denoising effectiveness of the DDAE was evaluated through a series of experiments. The study leveraged the synthetic dataset construction approach described earlier, where controlled combinations of source and noise signals were embedded within field-acquired surface recordings. By varying the frequency offset between the tool carrier signal and coherent noise components, the robustness of the DDAE under different spectral proximities could be quantified.

Figure 6.7 summarizes these results. It shows that the DDAE achieved its highest improvement in E_b/N_0 when the source and noise shared the same frequency, realizing an enhancement of 10.3 dB. As the separation increased to 3, 6 and 12 Hz, the absolute E_b/N_0 improvement decreased, though the DDAE continued to outperform the noisy input across all tested offsets.

This result illustrates an important contrast with traditional filtering approaches. Conventional bandpass filters rely on non-overlapping

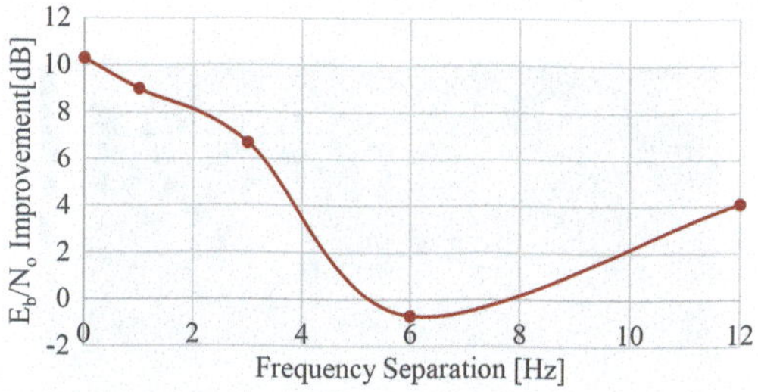

Fig. 6.7 Source frequency separation's impact on the DDAE denoising performance. The DDAE denoising performance improved when the source signal and the noise were closer in frequency. The best performance occurred at 0 Hz of separation, and it improved E_b/N_0 by 10.3 dB (Urdaneta et al. 2023)

frequency bands to isolate signals from interference. When signal and noise share the same frequency content, such filters fail to suppress the noise without also distorting the signal. In contrast, the DDAE learned spatial and temporal patterns in the multistake inputs to identify the coherent telemetry waveform even when its frequency content matched that of the noise.

Additional insight is provided by Fig. 6.8, which plots the frequency spectra of the DDAE output across the tested frequency separations. These plots demonstrate that the DDAE effectively reduced the spectral energy of the noise components while preserving the primary telemetry band, regardless of the offset. This capability to mitigate overlapping-frequency noise underscores the potential of deep learning denoisers to outperform traditional methods in scenarios typical of EM telemetry, where interference often resides within the same narrow operational band.

The improvement is further supported by constellation plots shown in Fig. 6.9. These plots visualize the demodulated QPSK symbols before and after DDAE processing, annotated with circles scaled by standard deviation to indicate E_b/N_0. At 0 Hz separation, the DDAE reduced the spread of the constellation points, directly translating to a substantial improvement in demodulation quality. Similar trends were evident at higher separations, though with diminishing gains as the offset increased.

Together, these experiments confirm that the DDAE is effective when source and noise signals occupy overlapping frequencies. This characteristic is relevant for downhole telemetry scenarios in conductive formations, where induced EM noise appears at the same low frequencies used for telemetry. It also establishes a key advantage over traditional frequency-domain filters, positioning learned denoising models as a solution for improving the reliability of EM telemetry under challenging spectral conditions.

6.4.2 Impact of Source Amplitude on DDAE

The experiments also investigated how the amplitude of the downhole source signal influenced the denoising performance of the DDAE. This is a practical consideration for EM telemetry applications, since downhole source strength can vary significantly with tool design, formation conditions, and operational constraints.

To examine this effect, the synthetic dataset methodology embedded downhole tool signals of different amplitudes into the same background of field-acquired surface noise and coherent interference. The amplitudes tested ranged from 10 mV, representing a severely attenuated source, up to 158 mV, which approximated a stronger transmission scenario. Each of these test cases was then processed through the DDAE trained as previously described.

Figure 6.10 presents the summary of E_b/N_0 improvements achieved by the DDAE across these different source amplitudes. The results demonstrate that higher source amplitudes generally corresponded to better denoising outcomes. For instance, at the highest tested source amplitude of 158 mV, the DDAE improved the E_b/N_0 by 4.1 dB, elevating the input mixture from 23.7 dB to 27.8 dB. As the amplitude decreased, the improvement tapered, with smaller gains at moderate amplitudes.

A different result occurred at the lowest tested amplitude of 10 mV. In this case, the raw mixture was so noisy that demodulation was not feasible,

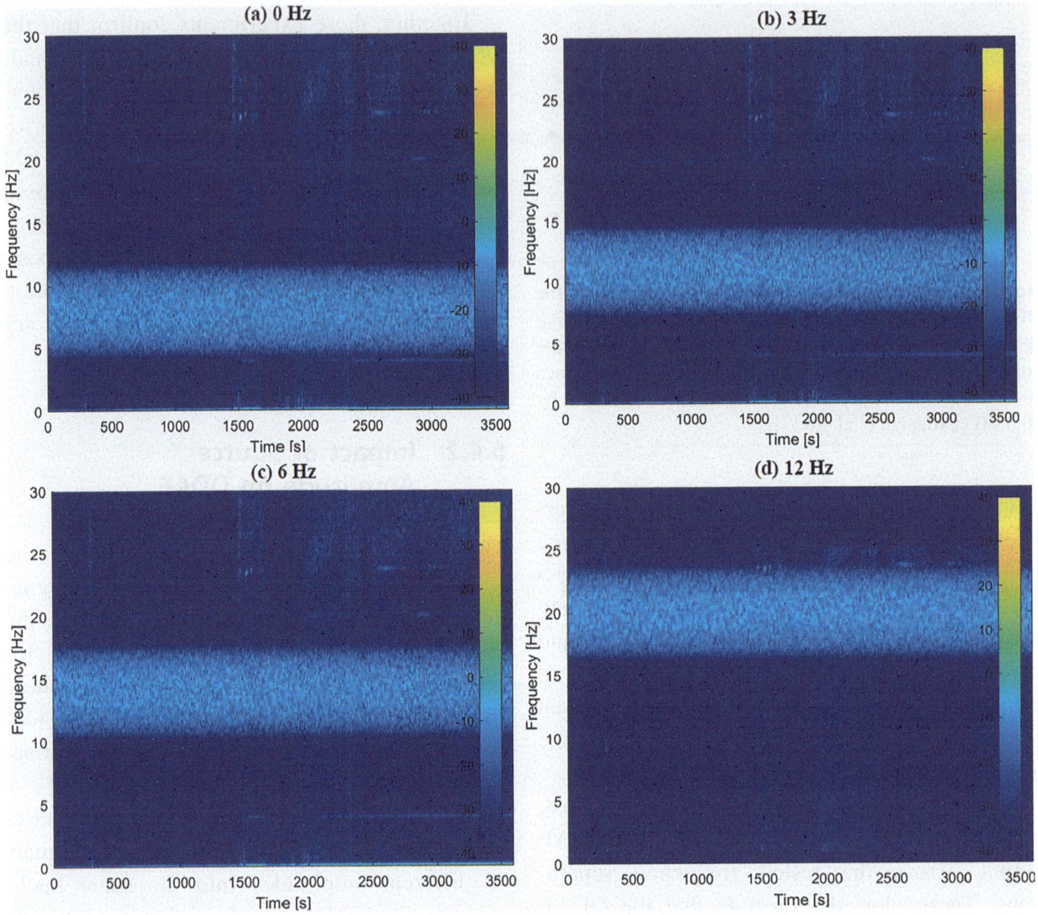

Fig. 6.8 Frequency spectra of different frequency separations, after applying DDAE denoising, confirm that noise has been removed from the input mixture (Urdaneta et al. 2023)

producing an undefined input E_b/N_0. However, after DDAE processing, the demodulated output achieved a measurable E_b/N_0 of 1.27 dB. This suggests that even at very low source strengths, where the mixture is below the minimum quality required for telemetry decoding, the DDAE was able to recover enough of the coherent signal structure to enable subsequent demodulation. This finding shows the potential of DL denoisers to extend the operational envelope of EM telemetry systems into regimes that would otherwise be infeasible.

The mechanisms of this performance are illustrated by the frequency domain plots shown in Fig. 6.11. These spectra demonstrate how the DDAE selectively preserved the telemetry carrier band across varying source amplitudes, while reducing broader noise components. At the lowest amplitude, the recovered spectral peak is modest, yet it remains resolved to support improved decoding compared to the original mixture.

These results indicate that the DDAE is effective when the downhole source maintains a reasonable amplitude relative to noise but also provides benefits even in degraded conditions. This reliability is key for field applications, where downhole EM transmission power may be constrained by hardware limitations, or where significant attenuation is introduced by formation conductivity and well geometry.

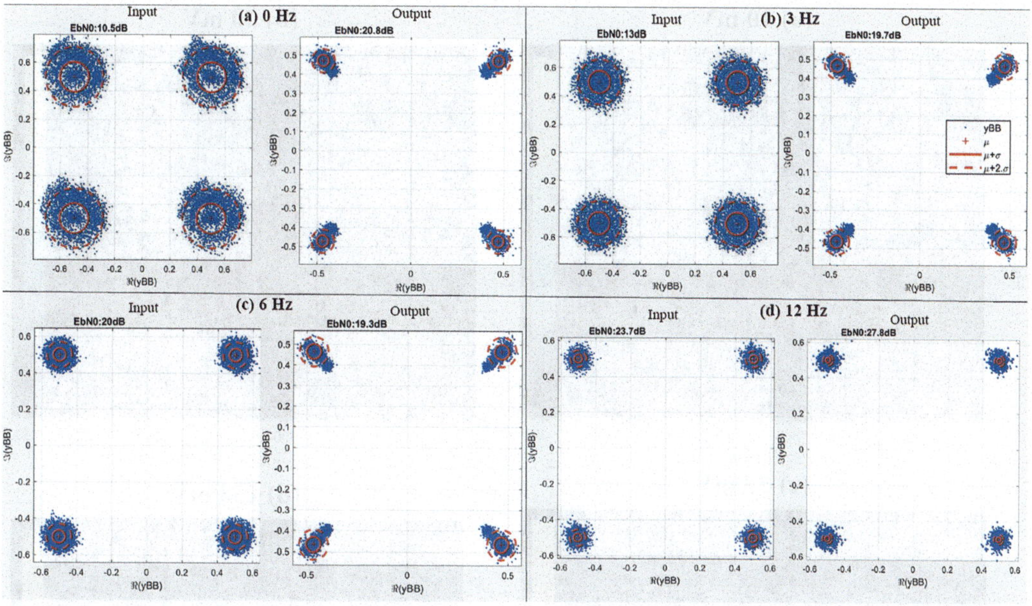

Fig. 6.9 Constellation plots of different frequency separations which confirm that the DDAE can filter noise at various frequency separations. The noisy input constellation plots have larger standard deviations representing a smaller E_b/N_0 value, while the denoised constellation plots have smaller standard deviations representing a higher Eb/N0 value (Urdaneta et al. 2023)

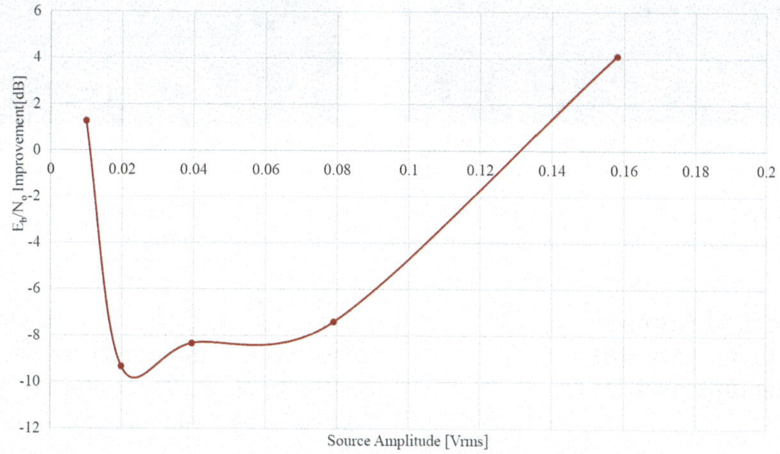

Fig. 6.10 Source amplitude's impact on the DDAE denoising performance. The denoising performance deteriorated with smaller source amplitudes. However, there is an exception for cases where the mixture is too noisy to be demodulated. In this case, the denoising performance improved even at smaller source amplitudes (Urdaneta et al. 2023)

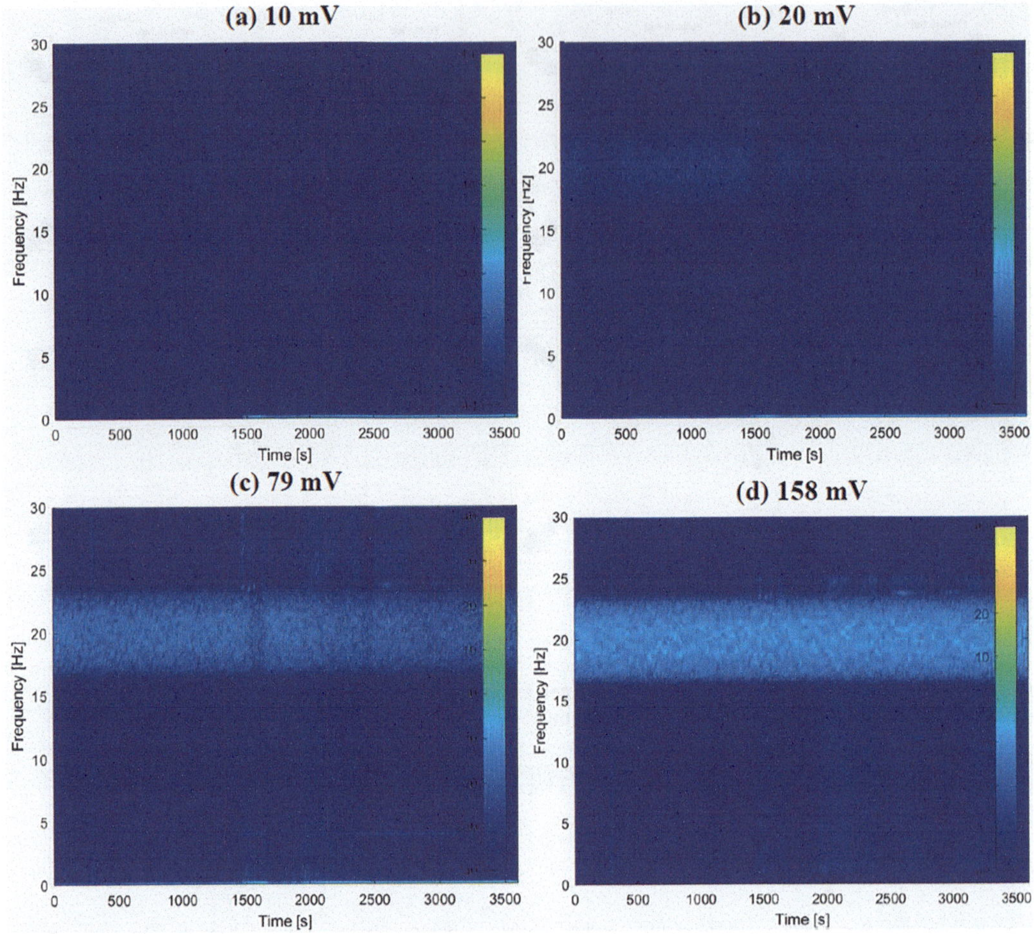

Fig. 6.11 Frequency spectra of different source amplitudes, after applying DDAE denoising, which confirm that the denoising performance deteriorated with smaller source amplitudes (Urdaneta et al. 2023)

6.4.3 Impact of Number of Stakes and Stake Placement on DDAE

The experiments also examined how the configuration of the surface stake array impacted the DDAE's ability to enhance telemetry signal quality. This is a key aspect of EM telemetry deployment, as the number of stakes, their distance from the wellhead, and their spatial orientation all affect how well the surface system captures the voltage differentials induced by downhole transmissions.

To quantify these effects, the field dataset included measurements from eight stakes positioned at varying distances and resistances relative to the wellhead. The denoising experiments altered the subset of stakes fed to the DDAE, assessing performance from single-stake cases up to the full eight-stake configuration.

The results, summarized in Fig. 6.12, show that increasing the number of stakes from one to two delivered a substantial improvement in denoising performance. Specifically, with two stakes, the DDAE improved E_b/N_0 by approximately 4 dB compared to the noisy mixture. However, adding more stakes beyond this threshold did not yield consistent further improvements. In some combinations, the inclusion of additional stakes even slightly degraded performance.

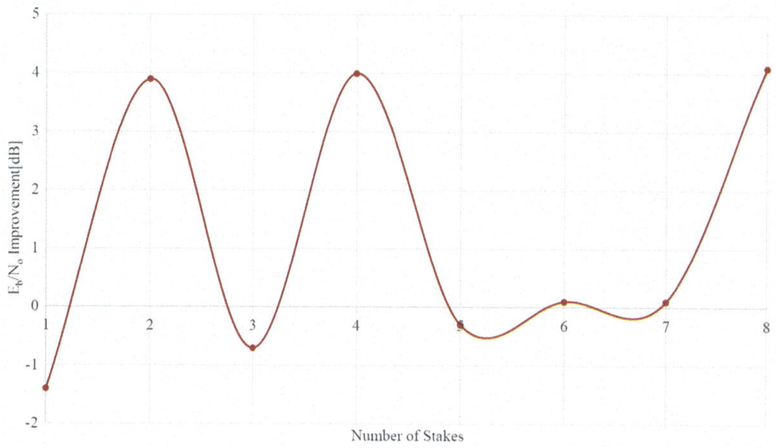

Fig. 6.12 Impact of the number of stakes on the DDAE denoising performance. The DDAE denoising performance improved when increasing the number of stakes from one to two, but it did not continue to increase with more than two stakes (Urdaneta et al. 2023)

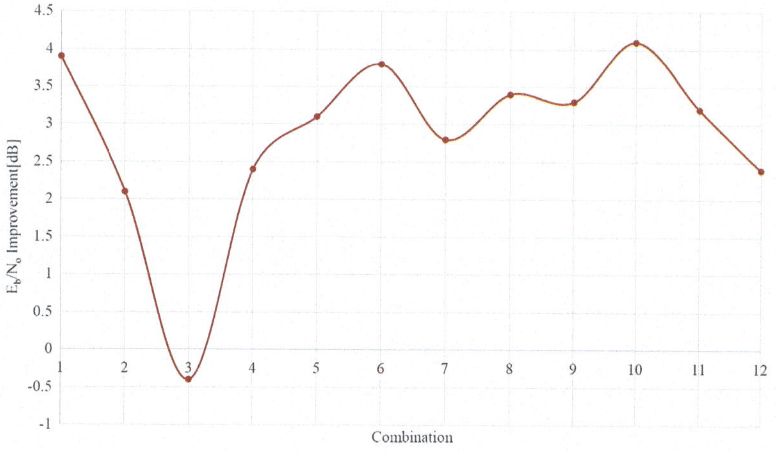

Fig. 6.13 Impact of a two-stake array's length and location on the DDAE denoising performance. The best performing case, Combination 10, was an orthogonal array (Urdaneta et al. 2023)

This effect can be attributed to variability in individual stake characteristics, such as ground resistance and position relative to local EM interference. For instance, stake 3 exhibited high resistance, while stake 5 had notably low resistance. Including these stakes altered the input data distribution in ways that did not uniformly enhance the DDAE's ability to extract coherent downhole signals.

To further explore spatial effects, the experiments also tested specific two-stake array arrangements. Figure 6.13 highlights that arrays configured with orthogonal orientation, meaning the stakes are positioned to capture voltage differences along perpendicular horizontal directions, performed better than parallel configurations. The best performance was observed in combination 10, an orthogonal layout, which maximized the DDAE's ability to reconstruct the downhole telemetry signal while suppressing noise.

Single-stake experiments provided additional insight into deployment constraints. As shown in Fig. 6.14, DDAE performance with only one stake was consistently poor across all positions tested. Figure 6.3 illustrates the diverse spatial distribution and resistances of the stakes in the dataset. Regardless of distance or resistance, no single stake provided enough spatial sampling of the EM field to allow the DDAE to reliably disentangle the downhole signal from ambient noise. In fact, the E_b/N_0 for single-stake outputs was frequently worse than the original input mixture.

These findings indicate that while the DDAE does not require a large number of stakes to achieve effective denoising, it does depend on

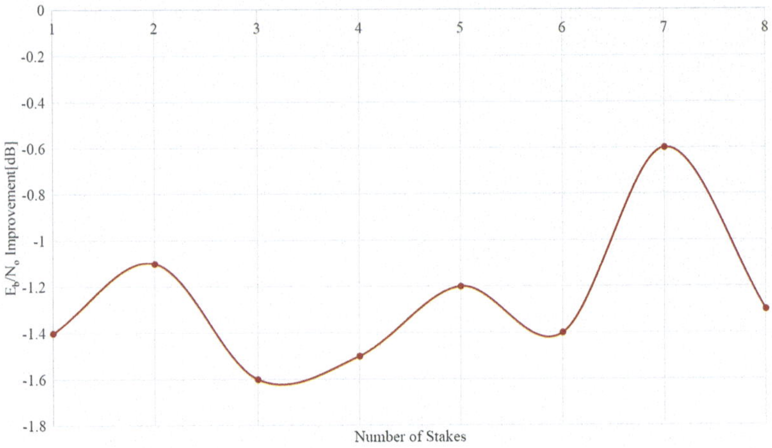

Fig. 6.14 Impact of single-stake position on the DDAE denoising performance. The DDAE denoising performance with a single stake was poor in all positions tested. This means that the DDAE is not suitable to be used with only one stake, regardless of its position (Urdaneta et al. 2023)

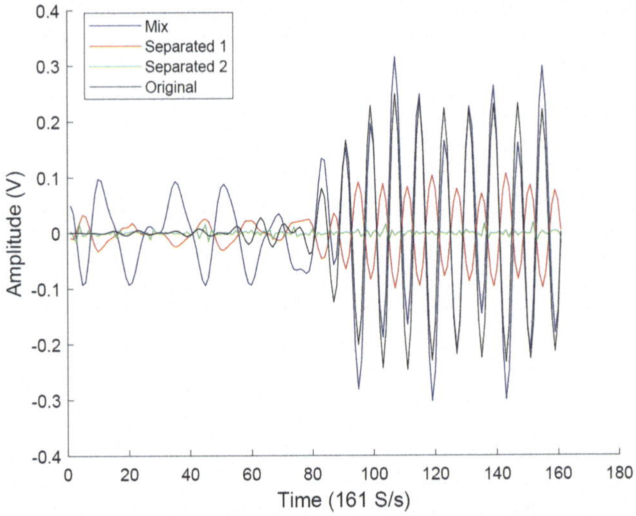

Fig. 6.15 Time series view of source separation with DPRNN. The acquired surface signal (shown as Mix) is separated into two waveforms: the downhole tool signal (shown as Separated 1) and the noise (shown as Separated 2). DPRNN is able to separate the downhole tool signal from the mixture, but the amplitude is still smaller than the original, which means there is room for improvement (Urdaneta et al. 2023)

having at least two well-placed stakes to establish the spatial diversity needed for learning and reconstructing the downhole signal. Moreover, the orientation of the stakes plays a meaningful role, suggesting that practical deployment guidelines should prioritize orthogonal layouts to maximize field capture.

6.4.4 DPRNN Separation Performance: Time Series and Spectral Insights

The experiments also evaluated the performance of the DPRNN architecture, which differs fundamentally from the autoencoder-based approaches. Unlike the DDAE and DDUL that focus on denoising by reconstructing a single cleaned signal, the DPRNN performs explicit source separation. It learns to produce two parallel output streams, one aimed at reconstructing the downhole telemetry signal, and another isolating the combined noise and interference components.

Figure 6.15 provides a time series view of this separation process. In the case shown, the downhole tool transmitted a synthetic source signal at 20 Hz carrier frequency and 8 bps with QPSK modulation. The acquired surface mixture clearly displayed significant amplitude fluctuations caused by interference and rig noise. When processed by the DPRNN, the output is separated into two distinct traces. The trace labeled "separated 1" closely tracked the shape of the original downhole signal, restoring amplitude stability lost in the mixed input. Meanwhile, the "separated 2" trace captured residual noise components that had previously masked the telemetry waveform.

Although the DPRNN restored much of the underlying signal structure, it was noted that the amplitude of the recovered signal was smaller than the original downhole waveform. This suggests that while the DPRNN isolates the telemetry signal from noise, it also compresses its range, indicating there remains room to further calibrate the recovered signal to match true amplitude levels.

The spectral view complements these time-domain findings. Figure 6.16 illustrates the frequency spectra and constellation plots for the DPRNN output under the same test conditions. The spectra show that the DPRNN was able to substantially reduce the power of the noise across the passband, concentrating energy around the expected telemetry carrier. This sharpening of the spectral content aligns with improved demodulation performance.

The constellation plots reinforce this improvement. In the noisy input, the constellation is dispersed, with large standard deviation circles indicating low E_b/N_0. After DPRNN separation, the constellation points tighten. The E_b/N_0 improved by approximately 3.3 dB,

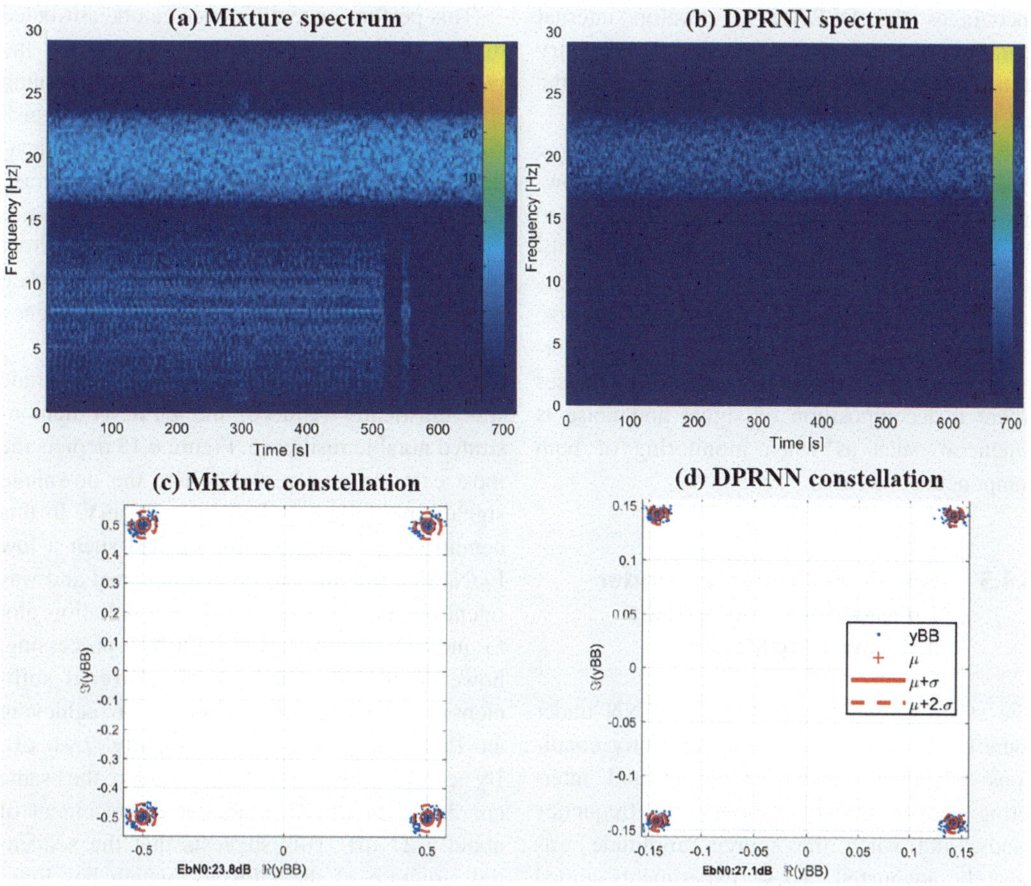

Fig. 6.16 Spectra and constellation plots of DPRNN optimized test cases with the best denoising performance and one stake. E_b/N_0 improved by 3.3 dB, taking the input mixture from 23.8 to 27.1 dB (Urdaneta et al. 2023)

increasing from 23.8 dB in the input mixture to 27.1 dB after separation. This represents an improvement, reducing symbol spread and enabling more reliable decoding.

These results show several characteristics of the DPRNN. First, its sequential modeling capability allows it to effectively disentangle temporally structured downhole signals from overlapping rig and formation noise. The recurrent design, with alternating intra-chunk and inter-chunk processing, captures both local signal coherence and long-range dependencies across the time sequence. This property is well suited for telemetry streams that encode information over extended symbol periods.

Second, the dual-path output structure ensures that the network is directly penalized if it fails to cleanly separate signal from noise. This design encourages the DPRNN to develop internal representations that distinctly allocate telemetry features to one channel and interference to the other.

The DPRNN demonstrates reliable source separation behavior on challenging low-frequency telemetry data. However, compared to the DDAE, it does exhibit differences in amplitude fidelity and more modest improvements in certain overlapping frequency scenarios. However, its time series and spectral characteristics suggest it is a suitable alternative in cases where a decomposition of signal and noise is beneficial, such as when monitoring of both components is desired.

6.4.5 DPRNN Performance Under Overlapping Frequencies and Low Amplitudes

The study further assessed the DPRNN under some of the most challenging telemetry conditions when the downhole signal and interfering noise shared overlapping frequency bands, and when the source amplitude was severely attenuated. These experiments aimed to clarify the relative strengths of the DPRNN compared to the DDAE, which had demonstrated distinct advantages under frequency overlap but more sensitivity to low amplitude conditions.

Figure 6.17 illustrates the scenario of complete frequency overlap, where both the downhole source signal and the coherent noise were set at 8 Hz. In this test, the DPRNN was able to improve the E_b/N_0 by approximately 0.94 dB, elevating the metric from 23.7 dB in the mixture to 24.6 dB after separation. This result shows that the DPRNN still provided a measurable improvement even when the spectral content of signal and noise were fully entangled, outperforming the traditional bandpass filter, which actually degraded E_b/N_0 by about 0.72 dB. However, it also confirmed that in these overlapping frequency conditions, the DDAE offered a substantially larger benefit, with an E_b/N_0 improvement of 10.3 dB in similar tests.

This performance difference can be attributed to the distinct operational principles of the two models. The DDAE relies on learning spatial patterns and local features that distinguish coherent downhole signals from noise, independent of strict frequency separation. In contrast, the DPRNN emphasizes temporal sequence modeling and separation through recurrent mechanisms, which may be less effective when both sources persist with identical frequency content over time.

On the other hand, when source amplitude was drastically reduced, the DPRNN demonstrated notable resilience. Figure 6.18 depicts the most extreme case tested, where the downhole signal amplitude was limited to 10 mV. In this condition, the original mixture had such a low E_b/N_0 that it could not be demodulated and was intentionally left blank in the constellation plot to indicate failure. After DPRNN processing, however, the separated signal recovered sufficient integrity to enable demodulation, achieving an E_b/N_0 improvement of roughly 7.91 dB. By comparison, the DDAE under the same conditions produced a smaller enhancement of about 1.27 dB. This suggests that the sequential structure of the DPRNN, which has long-range temporal correlations, is particularly effective at amplifying weak but consistent telemetry patterns buried in noise.

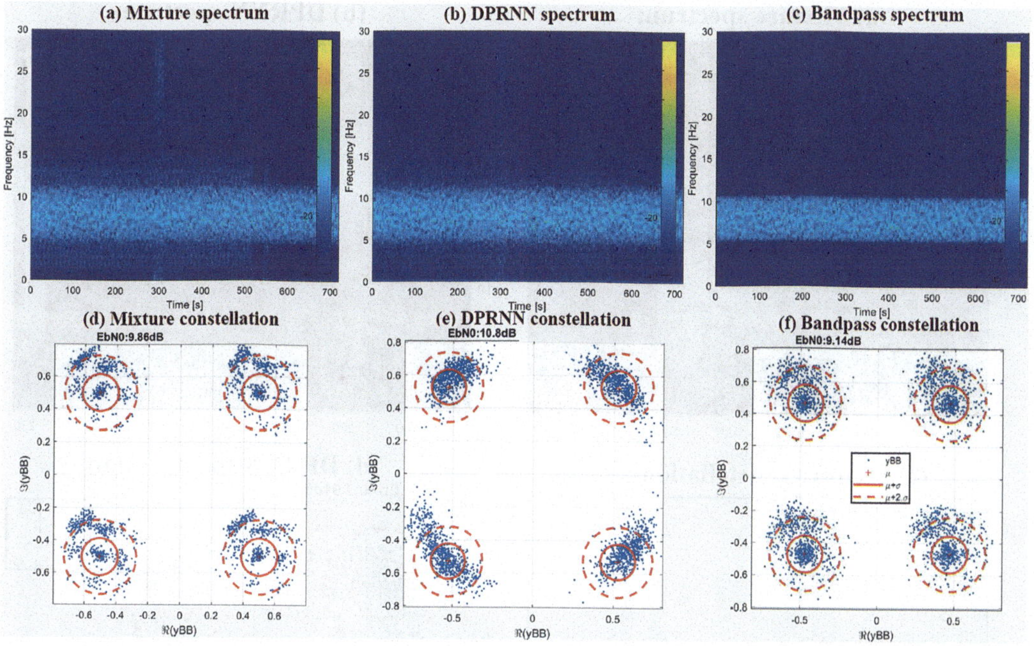

Fig. 6.17 Source frequency separation's impact on the DPRNN denoising performance, and how it compares with a traditional filter. DPRNN improved E_b/N_0 by 0.94 dB, which indicates that the DDAE is better suited for cases with overlapping frequencies because it provided an improvement of 10.3 dB. Both DPRNN and DDAE were better than a traditional bandpass filter, which worsened E_b/N_0 by 0.72 dB (Urdaneta et al. 2023)

Additionally, the experiments revisited the impact of the number of stakes for the DPRNN. As shown earlier with the DDAE, increasing the number of stakes from one to two provided clear benefits, but further increases did not yield substantial additional improvements. In the case of the DPRNN, depicted in Fig. 6.19, increasing the number of stakes beyond one did not significantly change the separation performance. The model achieved an E_b/N_0 improvement of approximately 3.1 dB even with only one stake, which was superior to the DDAE's performance under the same condition where E_b/N_0 actually worsened by about −1.4 dB. This result shows the operational advantage of the DPRNN for sites where only a limited number of stakes can be deployed.

These findings reinforce observations for model selection for EM telemetry enhancement. The DDAE shows superior noise discrimination under overlapping frequency conditions, using feature learning across multiple stakes. On the other hand, the DPRNN excels in situations with very weak signals or minimal stake configurations, using its temporal separation to recover the telemetry waveform from deep within the noise floor.

These complementary behaviors indicate that a hybrid strategy, or a conditional deployment where model choice is informed by expected noise conditions, could potentially offer the most reliable approach to improving telemetry reliability. Future studies might explore adaptive frameworks that switch or blend architectures depending on real-time assessments of frequency overlap and signal strength, taking advantage of both models.

6.5 Conclusion

This chapter presented a detailed investigation into the use of DL methods for enhancing the reliability of EM telemetry signals acquired from downhole tools. Addressing the operational challenge posed by SNR environments,

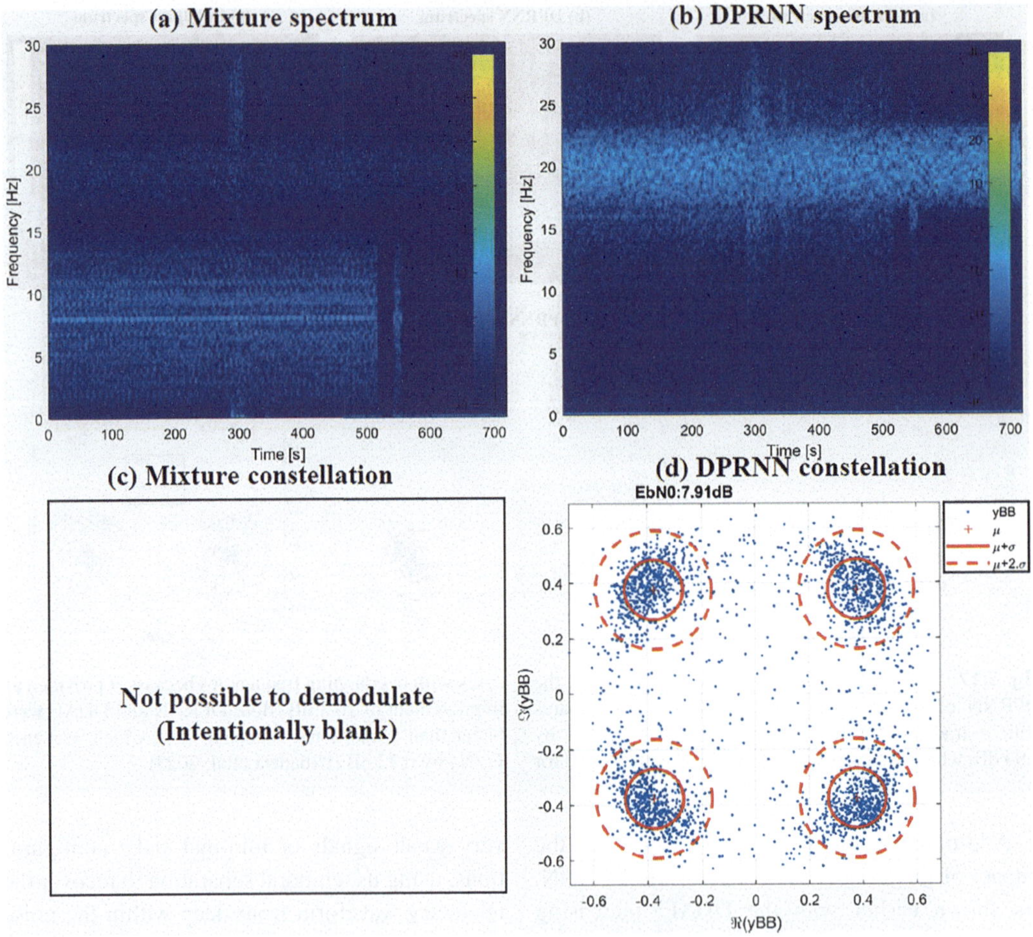

Fig. 6.18 Spectra and constellation plots for impact of source amplitude on DPRNN denoising performance. The mixture constellation E_b/N_0 was so low that it was not possible to demodulate it and intentionally shown blank. However, it was possible to demodulate the separated output. The DPRNN is better suited for cases with severe source amplitude attenuation because it provided an Eb/No improvement of 7.91 dB while DPRNN improved E_b/N_0 by 1.27 dB (Urdaneta et al. 2023)

the study focused on two DL frameworks: DDAE and DPRNN. These models were evaluated across a range of scenarios using a dataset that combined actual wellsite recordings from multiple stakes with synthetically injected downhole source signals to simulate clean telemetry references.

The experiments demonstrated that both architectures improved telemetry SNR over traditional filtering approaches, but each exhibited specific strengths depending on the characteristics of the telemetry environment. The DDAE, using its layered autoencoder structure and training phases that included both supervised initialization with synthetic labels and unsupervised fine-tuning on field data, excelled in scenarios where the downhole signal and noise had overlapping frequencies. It achieved an improvement in E_b/N_0 of approximately 10.3 dB under conditions of complete frequency overlap, outperforming traditional bandpass filters which degraded the SNR.

However, the DDAE showed limitations when confronted with very low signal amplitudes. In cases where the source signal was reduced to 10 mV, the DDAE still provided a modest

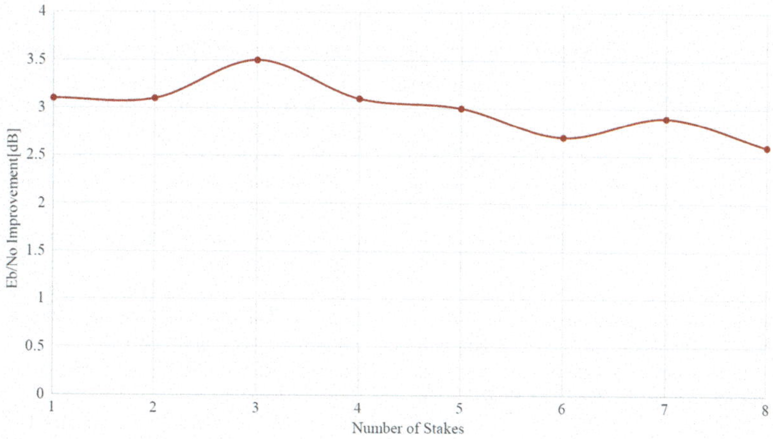

Fig. 6.19 Impact of number of stakes on DPRNN denoising performance. Increasing the number of stakes beyond one did not meaningfully improve the DPRNN denoising performance. The DPRNN is better suited for cases with only one stake because it provided an E_b/N_0 improvement of 3.1 dB while DDAE worsened E_b/N_0 by −1.4 dB (Urdaneta et al. 2023)

improvement of about 1.27 dB, but this fell short of restoring demodulation capability, showing the sensitivity of this architecture to weak telemetry signals.

In contrast, the DPRNN, structured around a segmentation and recurrent block processing scheme designed to capture both local and global temporal dependencies, proved particularly effective under severe amplitude attenuation. When tested under the same 10 mV scenario, the DPRNN recovered an E_b/N_0 improvement of approximately 7.91 dB, sufficient to restore signal demodulation where the original mixture could not be processed at all. This shows the model's ability to extract faint consistent patterns across time, more effectively than the DDAE.

The models also differed in their dependence on the number and arrangement of stakes. While the DDAE required at least two stakes to achieve significant improvements and showed sensitivity to stake array geometry, performing best with orthogonal configurations, the DPRNN maintained reliable performance even when operating with data from only a single stake. This characteristic makes the DPRNN valuable in operational contexts where constraints limit the deployment of stake arrays.

These findings provide a basis for selecting DL approaches tailored to telemetry challenges. The DDAE is suited for cases involving overlapping frequency environments and where multiple stakes can be deployed. The DPRNN offers advantages in hardware deployments and under conditions of signal attenuation. One promising area of future work involves integrating these models into hybrid or conditional pipelines that adaptively switch between architectures based on real-time assessments of downhole and surface noise characteristics.

In conclusion, the application of DL to EM telemetry noise attenuation and source separation offers a pathway to extend the operational envelope of telemetry systems. By enabling reliable signal extraction in noisy environments, these methods support more consistent real-time data availability from downhole tools, improving the ability to monitor and steer drilling operations effectively.

References

Abbaszadeh Shahri, A., Shan, C., Larsson, S., & Johansson, F. (2024). Normalizing large scale sensor based MWD data: An automated method toward a unified database. Sensors, 24(4), 1209.

Al-Fakih, A., Koeshidayatullah, A., Mukerji, T., & Kaka, S. I. (2024). Enhanced anomaly detection in well log data through the application of ensemble GANs. arXiv preprint arXiv:2411.19875.

Alsaihati, A., & Alotaibi, B. (2022). Determining severity of lateral and torsional downhole vibrations while drilling surface holes using three machine learning techniques. SPE Journal, 27(3), 1493–1503.

Altindal, M. C., Nivlet, P., Tabib, M., Rasheed, A., Kristiansen, T. G., & Khosravanian, R. (2024). Anomaly detection in multivariate time series of drilling data. Geoenergy Science and Engineering, 237, 212778.

Aoun, M. A. B., & Madarász, T. (2022). Applying machine learning to predict the rate of penetration for geothermal drilling located in the Utah FORGE site. Energies, 15(12), 4288.

Atashnezhad, A., Akhtarmanesh, S., Hareland, G., & Al Dushaishi, M. (2021). Developing a drilling optimization system for improved overall rate of penetration in geothermal wells. In ARMA US Rock Mechanics/Geomechanics Symposium. ARMA.

Bergstra, J., & Bengio, Y. (2012). Random search for hyper-parameter optimization. Journal of Machine Learning Research, 13, 281–305.

Biggio, L., Bendinelli, T., Kulkarni, C., & Fink, O. (2023). Ageing-aware battery discharge prediction with deep learning. Applied Energy, 346, 121229.

Carpenter, C. (2015). Review of downhole wireless communication techniques. Journal of Petroleum Technology, 67(3), 120–122.

Champion, B. P. (2016). Wireless reservoir monitoring reducing reservoir uncertainty during Barents Sea appraisal: A case history for Norvarg. In SPE Norway Subsurface Conference. SPE.

Ding, Z., Chen, Y., Wang, H., Wang, X., Zhang, W., & Zhang, Y. (2025). TIformer: A transformer-based framework for time-series forecasting with missing data. In Australasian Database Conference (pp. 71–84). Springer.

Dinh, N. M. (2020). Early kick detection using data-driven Bayesian network: Model development and experimental testing (Doctoral dissertation, Memorial University of Newfoundland).

Du, S., Huang, C., Ma, X., & Fan, H. (2024). A review of data-driven intelligent monitoring for geological drilling processes. Processes, 12(11), 2478.

Gidh, Y., Purwanto, A., & Smith Bits. (2012). Artificial neural network drilling parameter optimization system improves ROP by predicting and managing bit wear. In SPE Intelligent Energy International Conference and Exhibition. SPE.

Girase, H., Agarwal, N., Choi, C., & Mangalam, K. (2023). Latency matters: Real-time action forecasting transformer. In Proceedings of the IEEE/CVF Conference on Computer Vision and Pattern Recognition (pp. 18759–18769).

Gwynn, M., Allis, R., Hardwick, C., Jones, C., Nielsen, P., & Hurlbut, W. (2018). Rock properties of FORGE Well 58–32, Milford, Utah. Transactions, Geothermal Resources Council.

Habashy, T., & Anderson, B. I. (1991). Reconciling differences in depth of investigation between 2-MHz phase shift and attenuation resistivity measurements. In SPWLA Annual Logging Symposium. SPWLA.

Hegde, C., Soares, C., & Gray, K. (2018). Rate of penetration modeling using hybrid models: Deterministic and machine learning. In Unconventional Resources Technology Conference (pp. 3220–3238).

Hermawan Manuaba, I. B. G., Aljishi, M., Van Steene, M., & Dolan, J. (2024). Logging-while-drilling laterolog vs. electromagnetic propagation measurements: Which is telling the true resistivity? SPE Journal, 29(8), 4000–4013.

Hyndman, R. J. (2018). Forecasting: Principles and practice. OTexts.

Iglesias, G., Talavera, E., González-Prieto, Á., Mozo, A., & Gómez-Canaval, S. (2023). Data augmentation techniques in time series domain: A survey and taxonomy. Neural Computing and Applications, 35(14), 10123–10145.

Islam, N., Vijapurapu, R., Jones, M., McLennan, J., Moore, J., Rickard, W., Balamir, O., & Vagnetti, R. (2018). Application of mechanical specific energy and at-the-bit measurements for geothermal drilling applications. In ARMA US Rock Mechanics/Geomechanics Symposium. ARMA.

Jannin, G., Chen, J., DePavia, L. E., Sun, L., & Schwartz, M. (2017). Deep electrode: A game-changing technology for electromagnetic telemetry. In SEG International Exposition and Annual Meeting. SEG.

Keman, L. (2016). Adaptive noise cancellation for electromagnetic-while-drilling system. In International Conference on Information Science and Control Engineering (pp. 1253–1256). IEEE.

Kingma, D. P., & Welling, M. (2019). An introduction to variational autoencoders. Foundations and Trends in Machine Learning, 12(4), 307–392.

Koeshidayatullah, A., Al-Fakih, A., & Kaka, S. I. (2024). Leveraging time-series foundation model for subsurface well logs prediction and anomaly detection. arXiv preprint arXiv:2412.05681.

Kong, Z., Xu, Y., Zhang, Z., & Luo, B. (2023). Diffusion models for time series applications: A survey. arXiv preprint arXiv:2305.00624.

Luo, Y., Chen, Z., & Yoshioka, T. (2020). Dual-path RNN: Efficient long sequence modeling for time-domain single-channel speech separation. In ICASSP 2020 (pp. 46–50). IEEE.

Macune, D. T., Flanagan, W. D., Choi, E., & Marcellus, E. (2006). A compact compensated resistivity tool for logging while drilling. In SPE/IADC Drilling Conference and Exhibition. SPE.

Mikołajewska, E., Mikołajewski, D., Mikołajczyk, T., & Paczkowski, T. (2025). Generative AI in AI-based digital twins for fault diagnosis for predictive maintenance in Industry 4.0/5.0. Applied Sciences, 15(6), 3166.

Mora-de-León, L. P., Solís-Martín, D., Galán-Páez, J., & Borrego-Díaz, J. (2025). Text-conditioned diffusion-based synthetic data generation for turbine engine sensor analysis and RUL estimation. Machines, 13(5), 374.

Motahhari, H. R., Hareland, G., & James, J. A. (2010). Improved drilling efficiency technique using integrated PDM and PDC bit parameters. Journal of Canadian Petroleum Technology, 49(10), 45–52.

Okoh, C., Roy, R., Mehnen, J., & Redding, L. (2014). Overview of remaining useful life prediction techniques in through-life engineering services. Procedia CIRP, 16, 158–163.

Phelan, Z., Xing, P., Panja, P., Moore, J., & McLennan, J. (2022). Prediction of formation properties based on drilling data of wells at Utah FORGE site using machine learning. In ARMA US Rock Mechanics/Geomechanics Symposium. ARMA.

Rickard, W. M., McLennan, J., Islam, N., & Rivas, E. (2019). Mechanical specific energy analysis of the FORGE Utah well. In Proceedings of the 44th Workshop on Geothermal Reservoir Engineering. Stanford University.

Rykov, Y. (2024). Drilling and logging equipment reliability in a downhole vibration environment. Journal of Petroleum Technology.

Saad, O. M., & Chen, Y. (2020). Deep denoising autoencoder for seismic random noise attenuation. Geophysics, 85(4), V367–V376.

Salinas, D., Flunkert, V., Gasthaus, J., & Januschowski, T. (2020). DeepAR: Probabilistic forecasting with autoregressive recurrent networks. International Journal of Forecasting, 36(3), 1181–1191.

Shah, A. B., Wen, Y., Chen, J., Wu, X., & Fu, X. (2024). A deep learning approach to predicting remaining useful life for downhole drilling sensors using synthetic data generation. In Unconventional Resources Technology Conference. URTeC.

Srivastava, N., Hinton, G., Krizhevsky, A., Sutskever, I., & Salakhutdinov, R. (2014). Dropout: A simple way to prevent neural networks from overfitting. Journal of Machine Learning Research, 15, 1929–1958.

Sutharssan, T., Stoyanov, S., Bailey, C., & Yin, C. (2015). Prognostic and health management for engineering systems: A review of the data-driven approach and algorithms. Journal of Engineering, 2015(7), 215–222.

Urdaneta, C., Jeong, C., Wu, X., & Chen, J. (2024a). Deep learning method for improving rate of penetration prediction in drilling. SPE Journal.

Urdaneta, C., Jeong, C., Wu, X., & Chen, J. (2024b). Predicting coiled-tubing drilling dynamics using transformers. In Unconventional Resources Technology Conference. URTeC.

Urdaneta, C., Jeong, C., & Zheng, A. (2024). An artificial intelligence model to synthesize measurements while drilling sensors for coiled tubing drilling. In SPE/ICoTA Well Intervention Conference and Exhibition. SPE.

Urdaneta, C., Jeong, C., Wu, X., & Chen, J. (2025). Detecting resistivity tool malfunctions in coiled tubing drilling using generative artificial intelligence. In Unconventional Resources Technology Conference. URTeC.

Urdaneta, C., Jarrot, A., Wang, S., Wu, X., & Chen, J. (2023). Deep learning methods for improving electromagnetic telemetry signal-to-noise ratio. SPE Journal, 28(4), 1676–1689.

Vaswani, A., Shazeer, N., Parmar, N., Uszkoreit, J., Jones, L., Gomez, A. N., Kaiser, Ł., & Polosukhin, I. (2017). Attention is all you need. Advances in Neural Information Processing Systems.

Wang, J., Sajeev, S. K., Ozbayoglu, E., Baldino, S., Liu, Y., & Jing, H. (2023). Reducing NPT using a novel approach to real-time drilling data analysis. In SPE Annual Technical Conference and Exhibition. SPE.

Wang, H., Wang, Y., Ding, J., & Wang, J. (2024). Generative adversarial networks for synthetic bearing fault data augmentation: A review. IEEE Access, 12, 28456–28471.

Weiss, K., Khoshgoftaar, T. M., & Wang, D. (2016). A survey of transfer learning. Journal of Big Data, 3(1), 1–40.

Wen, R., Torkkola, K., Narayanaswamy, B., & Madeka, D. (2017). A multi-horizon quantile recurrent forecaster. arXiv preprint arXiv:1711.11053.

Wu, H., Xu, J., Wang, J., & Long, M. (2021). Autoformer: Decomposition transformers with auto-correlation for long-term series forecasting. Advances in Neural Information Processing Systems, 34, 22419–22430.

Yang, L., Wang, S., Chen, X., Saad, O. M., Chen, W., Oboue, Y. A. S. I., & Chen, Y. (2021). Unsupervised 3-D random noise attenuation using deep skip autoencoder. IEEE Transactions on Geoscience and Remote Sensing, 60, 1–16.

Zhou, H., Zhang, S., Peng, J., Zhang, S., Li, J., Xiong, H., & Zhang, W. (2021). Informer: Beyond efficient transformer for long sequence time-series forecasting. In Proceedings of the AAAI Conference on Artificial Intelligence (Vol. 35, pp. 11106–11115).

Zhuang, F., Qi, Z., Duan, K., Xi, D., Zhu, Y., Zhu, H., Xiong, H., & He, Q. (2021). A comprehensive survey on transfer learning. Proceedings of the IEEE, 109(1), 43–76.

7 Predicting Remaining Useful Life of Drilling Equipment

7.1 Background and Related Work

Accurate prediction of remaining useful life (RUL) is a core capability for predictive maintenance in drilling operations, where extreme temperatures, pressure cycles and mechanical stresses accelerate degradation while complete run-to-failure examples remain rare. Prior work in prognostics demonstrates that physics-informed features, sequence models such as LSTM and GRU, and careful validation by equipment are all important for reliable field performance. In this chapter we on an LSTM-based regression pipeline that combines domain-driven feature design with rigorous preprocessing and equipment-wise validation, and we show how that pipeline produces practical, conservative forecasts that are suitable for operational decision making.

7.1.1 Prognostics and Health Management in Drilling

Predictive maintenance through RUL estimation has emerged as a critical capability for managing downhole drilling equipment, where operational environments impose extreme thermal, mechanical, and chemical stresses that accelerate component degradation. Unlike surface equipment where scheduled maintenance intervals can be conservatively set, downhole tools operate in inaccessible locations where premature replacement incurs substantial tripping costs while delayed replacement risks catastrophic failures that can halt operations for days. This economic trade-off motivates accurate RUL forecasting that balances maintenance overhead against failure risk.

Traditional prognostics approaches in drilling rely on fixed maintenance schedules derived from manufacturer specifications and historical field experience (Okoh et al. 2014). However, these schedules cannot account for site-specific operational variations such as thermal cycling frequency, formation abrasiveness, and drilling fluid chemistry, leading to either excessive preventive maintenance or unexpected failures. Physics-based models that simulate component wear through stress-life relationships offer improved accuracy but require detailed material properties and loading histories that are rarely available for commercial drilling tools (Sutharssan et al. 2015).

7.2 Dataset Description and Pre-Processing

7.2.1 Dataset Pre-Processing

The primary telemetry for our experiments is drawn from the Utah FORGE collection, chosen because of its high temporal resolution and the presence of multiple sensors relevant to thermal

C. Urdaneta et al., *Deep Learning Applications for Drilling Performance: Forecasting, Diagnostics, Telemetry, and Predictive Maintenance*, Advances in Oil and Gas Exploration & Production,
https://doi.org/10.1007/978-3-032-26648-4_7

and mechanical stress. We limit our experiments to sensors commonly available in downhole telemetry streams and retain metadata such as tool identifiers and drilling phase annotations to enable stratified evaluation.

We standardize units, resample traces to a common cadence, and perform gap handling and outlier mitigation so downstream features operate on consistent inputs. Records with irrecoverable nulls are removed, and extreme outliers are flagged for manual inspection. To provide an explicit temporal supervision signal we add a life cycle index that increments by one per row, and from this index we compute per-trajectory RUL labels by subtracting the current life cycle from the trajectory peak. For consistency across examples, we cap RUL at a fixed horizon used in training, which aligns the learning problem and prevents undue influence from extremely long tails.

To encode physics knowledge in the input, we compute cumulative temperature exposure (CTE), as the running integral of casing or sensor temperature after baseline removal. CTE serves as an interpretable indicator of accumulated thermal stress and typically correlates negatively with RUL. In addition to CTE we compute mechanical proxies such as weight on bit and rotational speed, and rolling statistical descriptors such as multi-scale means, standard deviations and first differences to capture short-term volatility and longer-term trends.

We construct supervised examples with sliding windows of past observations that reflect the dominant degradation timescale. Each input window is paired with the RUL label at the window end. Window length, sampling cadence and forecast horizon are hyperparameters tuned to balance context coverage, training example count and computational cost.

7.2.2 Model Architecture and Training

Long short-term memory (LSTM) networks are well suited to prognostics because their gated cell state filters noisy telemetry and preserves long-range dependencies that are characteristic of cumulative degradation. The LSTM naturally captures non-linear temporal dynamics that emerge as equipment approaches end of life, and it integrates multiple signals across time without requiring manual window aggregation.

Our regressor uses stacked LSTM layers followed by dense projection heads. The configuration that produced the best validation performance in our experiments comprises two LSTM layers with 128 units each, dropout with rate 0.2 for regularization, a dense layer with 64 units and ReLU activation, and a single linear output neuron for RUL regression. This architecture balances modeling capacity with computational efficiency and is compact enough for practical retraining and fine-tuning in pilot deployments.

We train with the Adam optimizer, an initial learning rate of 1×10^{-3}, mini-batches of size 64, and mean squared error loss. Early stopping on validation loss with patience 10 prevents overfitting, and learning-rate reduction on plateau improves convergence stability. All experiments use deterministic random seeds for data splits, augmentation (when used downstream), and weight initialization.

7.2.3 Baseline Evaluation and Metrics

We compare the LSTM against common baselines: XGBoost and Random Forest tree ensembles trained on the same engineered features, a GRU-based recurrent model, a vanilla RNN, and a persistence heuristic. These baselines highlight the benefits of sequence modeling and provide operationally relevant comparisons.

We report normalized root mean squared error (RMSE), mean absolute percentage error (MAPE), and coefficient of determination R2. For robustness we report mean and standard deviation across equipment folds and provide per-site breakdowns to surface transportability issues.

To avoid temporal leakage, we partition by equipment using k-fold cross validation with $k = 5$. Each fold holds out entire equipment IDs for testing so that performance reflects generalization to unseen assets. Within each training fold we reserve the most recent portion of each equipment's history for validation to emulate forecasting on live streams.

7.3 Experimental Results and Analysis

7.3.1 Quantitative and Qualitative Performance

In the experiments, the proposed LSTM regressor achieved a normalized RMSE of approximately 7.6% and a MAPE of approximately 14.8% on the FORGE evaluation splits. Tree based baselines such as XGBoost and Random Forest produced higher error values, while classical recurrent variants such as GRU and vanilla RNN also showed larger prediction errors than the optimized LSTM configuration under the same evaluation conditions. These numerical differences indicate improved prediction accuracy for the LSTM model in scenarios where long range temporal dependencies are informative.

Per-well prediction traces show that the LSTM captures gradual declines and provides conservative late-life estimates, which is desirable for safety minded maintenance scheduling. Primary failure modes include abrupt mechanical events where telemetry exhibits sudden steps not reflected in historical patterns, and sensor faults that create localized artifacts. These cases motivate richer multi-modal inputs and improved anomaly handling in future work.

7.3.2 Feature Importance and Correlation Analysis

We inspected a comprehensive cross-correlation matrix (Fig. 7.1) to understand how engineered features relate to remaining useful life and to one another, because these relationships guide feature selection, regularization and interpretation. The matrix contains two obvious patterns we use to structure our interpretation.

First, the life cycle index is almost perfectly negatively correlated with RUL, which is expected because RUL is computed directly from life cycle. We therefore treat life cycle as a supervisory index rather than as a predictive covariate when reporting feature importance, and we exclude it from claims about causal signals so that downstream explanations focus on independently measured channels.

Second, among independently measured and engineered channels, temperature-derived indicators show the strongest and most consistent association with RUL. In particular, cumulative temperature exposure (CTE) exhibits the largest negative correlation with RUL (approximately −0.35), with exponential moving average (EMA) and raw temperature channels also showing sizeable negative associations (roughly −0.25 to −0.30). This pattern matches physical intuition: accumulated thermal stress is a plausible driver of progressive sensor and component degradation, and the monotonic nature of the cumulative measure explains its stronger relationship with remaining life.

By contrast, pressure features display much weaker correlations with RUL, typically near zero to slightly negative (on the order of −0.08 to −0.12). Simulated pump pressure and its running averages correlate with each other, and they provide useful contextual information about operational regimes, but they are not the primary RUL drivers in these experiments. Indicators that capture intermittent behavior, such as the probability-of-zero and zero-frequency features, show low to modest correlations with RUL on the aggregate, yet they are important for modeling rare, abrupt failures and for improving robustness to transient faults.

The matrix also reveals expected multicollinearity among engineered statistics. For example, EMA, rolling means and cumulative temperature exposure are mutually correlated, because they derive from the same underlying temperature channel at different time scales. Similarly, simulated pump pressure, average simulated pump pressure and related rolling

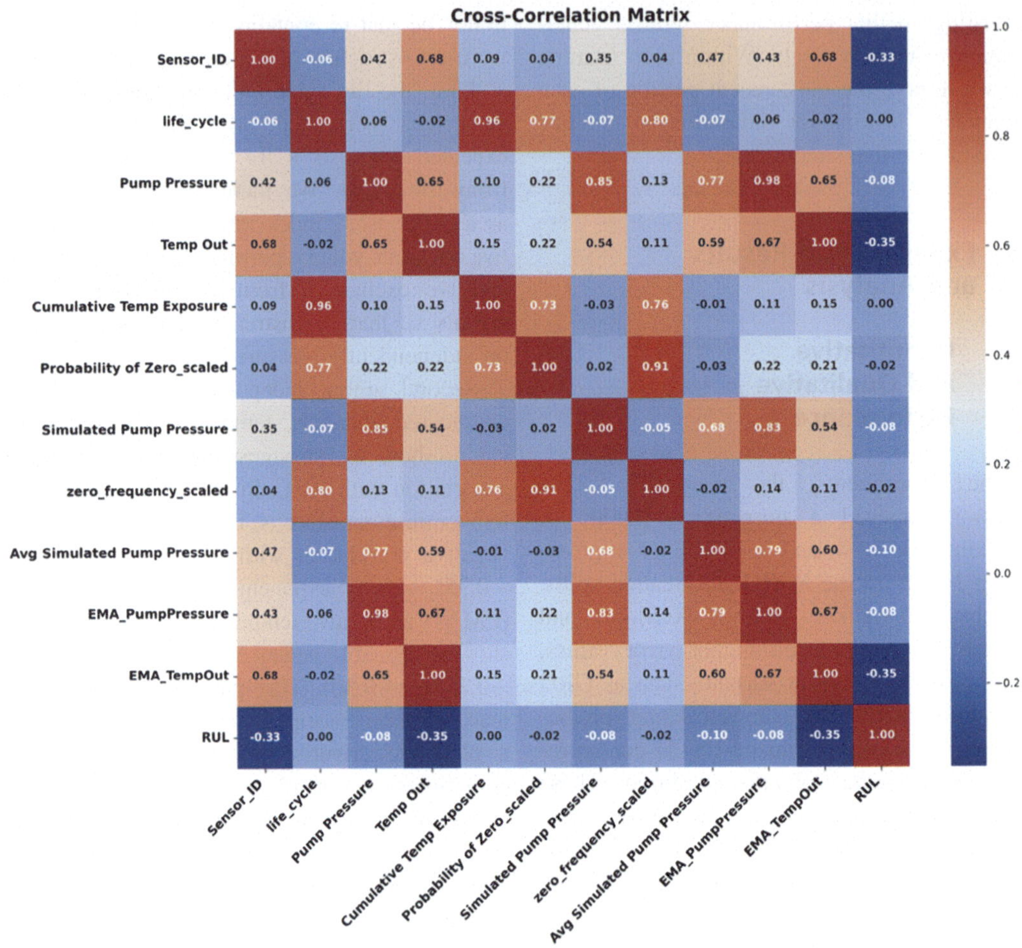

Fig. 7.1 Cross correlation matrix of features in UTAH FORGE dataset. The input features exhibit weak to moderate correlation with RUL with no single feature showing dominance over the prediction target

statistics form a correlated block. We account for these dependencies in modeling by using regularization, by preferring parsimonious projection heads in the network, and by relying on permutation-style importance or SHAP analyses rather than raw correlation magnitudes to assess each feature's marginal contribution to predictive performance.

7.3.3 Model Performance

Across the baselines, the LSTM achieves the best overall accuracy, with a normalized RMSE of 7.60% percent and a MAPE of 14.80%. XGBoost is the second-best performer with an RMSE of 8.28% and a MAPE of 17.30%, while GRU and classical recurrent models show intermediate performance and tree-based ensembles record the largest errors in our setup. These results indicate that recurrent sequence models that preserve long-range temporal dependencies are particularly effective for RUL forecasting when cumulative temperature features and synthetic run-to-failure examples are available, and that the LSTM configuration tuned in our experiments consistently yields the lowest point error across folds.

7.3.4 Hyperparameter Optimization

We conduct systematic hyperparameter searches to balance model capacity and generalization. The hyperparameter sweep identified four LSTM layers and a dense projection with 64 units as strong settings in our search grid, with dropout probabilities around 0.1 and an Adam optimizer at learning rate 0.001 giving the best validation behavior. The trained LSTM models converge reliably within 40 epochs under early stopping and reproducible seeding, and these tuned settings form the basis for the final reported experiments.

7.4 Conclusion

This chapter presented an LSTM-based framework for predicting remaining useful life of downhole drilling sensors, achieving 7.60% normalized RMSE and 14.80% MAPE on the Utah FORGE geothermal dataset. The methodology combines physics-informed feature engineering, synthetic data augmentation, and equipment-wise validation to address the dual challenges of sparse run-to-failure data and nonlinear degradation dynamics in high-temperature drilling environments.

Comparative evaluation against six baseline models demonstrated that the LSTM's explicit memory mechanisms provide substantial advantages over both classical machine learning methods (34–55% error reduction versus tree ensembles) and simplified recurrent variants (34% improvement versus GRU). Feature importance analysis confirmed that cumulative temperature exposure dominates RUL prediction with correlation –0.35 and permutation importance 0.42, validating the physics-informed design approach. Several limitations motivate future extensions. First, the framework operates exclusively on single-site geothermal data. Second, the current feature set may miss non-thermal failure modes associated with mechanical wear or chemical contamination. Third, point-estimate predictions lack explicit uncertainty quantification needed for risk-aware maintenance decisions.

The next chapter addresses the cross-site generalization challenge by harmonizing FORGE and offshore Volve datasets through extended synthetic augmentation and introducing a hybrid BiGRU-BiLSTM architecture that achieves 14.92% MAPE on combined data. Beyond the works in this book, future work should pursue multi-modal sensor fusion, uncertainty-aware architectures, and decision-aware training objectives that directly optimize maintenance cost reduction under operational constraints.

References

Abbaszadeh Shahri, A., Shan, C., Larsson, S., & Johansson, F. (2024). Normalizing large scale sensor based MWD data: An automated method toward a unified database. Sensors, 24(4), 1209.

Al-Fakih, A., Koeshidayatullah, A., Mukerji, T., & Kaka, S. I. (2024). Enhanced anomaly detection in well log data through the application of ensemble GANs. arXiv preprint arXiv:2411.19875.

Alsaihati, A., & Alotaibi, B. (2022). Determining severity of lateral and torsional downhole vibrations while drilling surface holes using three machine learning techniques. SPE Journal, 27(3), 1493–1503.

Altindal, M. C., Nivlet, P., Tabib, M., Rasheed, A., Kristiansen, T. G., & Khosravanian, R. (2024). Anomaly detection in multivariate time series of drilling data. Geoenergy Science and Engineering, 237, 212778.

Aoun, M. A. B., & Madarász, T. (2022). Applying machine learning to predict the rate of penetration for geothermal drilling located in the Utah FORGE site. Energies, 15(12), 4288.

Atashnezhad, A., Akhtarmanesh, S., Hareland, G., & Al Dushaishi, M. (2021). Developing a drilling optimization system for improved overall rate of penetration in geothermal wells. In ARMA US Rock Mechanics/Geomechanics Symposium. ARMA.

Bergstra, J., & Bengio, Y. (2012). Random search for hyper-parameter optimization. Journal of Machine Learning Research, 13, 281–305.

Biggio, L., Bendinelli, T., Kulkarni, C., & Fink, O. (2023). Ageing-aware battery discharge prediction with deep learning. Applied Energy, 346, 121229.

Carpenter, C. (2015). Review of downhole wireless communication techniques. Journal of Petroleum Technology, 67(3), 120–122.

Champion, B. P. (2016). Wireless reservoir monitoring reducing reservoir uncertainty during Barents Sea appraisal: A case history for Norvarg. In SPE Norway Subsurface Conference. SPE.

Ding, Z., Chen, Y., Wang, H., Wang, X., Zhang, W., & Zhang, Y. (2025). TIformer: A transformer-based framework for time-series forecasting with missing data. In Australasian Database Conference (pp. 71–84). Springer.

Dinh, N. M. (2020). Early kick detection using data-driven Bayesian network: Model development and experimental testing (Doctoral dissertation, Memorial University of Newfoundland).

Du, S., Huang, C., Ma, X., & Fan, H. (2024). A review of data-driven intelligent monitoring for geological drilling processes. Processes, 12(11), 2478.

Gidh, Y., Purwanto, A., & Smith Bits. (2012). Artificial neural network drilling parameter optimization system improves ROP by predicting and managing bit wear. In SPE Intelligent Energy International Conference and Exhibition. SPE.

Girase, H., Agarwal, N., Choi, C., & Mangalam, K. (2023). Latency matters: Real-time action forecasting transformer. In Proceedings of the IEEE/CVF Conference on Computer Vision and Pattern Recognition (pp. 18759–18769).

Gwynn, M., Allis, R., Hardwick, C., Jones, C., Nielsen, P., & Hurlbut, W. (2018). Rock properties of FORGE Well 58–32, Milford, Utah. Transactions, Geothermal Resources Council.

Habashy, T., & Anderson, B. I. (1991). Reconciling differences in depth of investigation between 2-MHz phase shift and attenuation resistivity measurements. In SPWLA Annual Logging Symposium. SPWLA.

Hegde, C., Soares, C., & Gray, K. (2018). Rate of penetration modeling using hybrid models: Deterministic and machine learning. In Unconventional Resources Technology Conference (pp. 3220–3238).

Hermawan Manuaba, I. B. G., Aljishi, M., Van Steene, M., & Dolan, J. (2024). Logging-while-drilling laterolog vs. electromagnetic propagation measurements: Which is telling the true resistivity? SPE Journal, 29(8), 4000–4013.

Hyndman, R. J. (2018). Forecasting: Principles and practice. OTexts.

Iglesias, G., Talavera, E., González-Prieto, Á., Mozo, A., & Gómez-Canaval, S. (2023). Data augmentation techniques in time series domain: A survey and taxonomy. Neural Computing and Applications, 35(14), 10123–10145.

Islam, N., Vijapurapu, R., Jones, M., McLennan, J., Moore, J., Rickard, W., Balamir, O., & Vagnetti, R. (2018). Application of mechanical specific energy and at-the-bit measurements for geothermal drilling applications. In ARMA US Rock Mechanics/Geomechanics Symposium. ARMA.

Jannin, G., Chen, J., DePavia, L. E., Sun, L., & Schwartz, M. (2017). Deep electrode: A game-changing technology for electromagnetic telemetry. In SEG International Exposition and Annual Meeting. SEG.

Keman, L. (2016). Adaptive noise cancellation for electromagnetic-while-drilling system. In International Conference on Information Science and Control Engineering (pp. 1253–1256). IEEE.

Kingma, D. P., & Welling, M. (2019). An introduction to variational autoencoders. Foundations and Trends in Machine Learning, 12(4), 307–392.

Koeshidayatullah, A., Al-Fakih, A., & Kaka, S. I. (2024). Leveraging time-series foundation model for subsurface well logs prediction and anomaly detection. arXiv preprint arXiv:2412.05681.

Kong, Z., Xu, Y., Zhang, Z., & Luo, B. (2023). Diffusion models for time series applications: A survey. arXiv preprint arXiv:2305.00624.

Luo, Y., Chen, Z., & Yoshioka, T. (2020). Dual-path RNN: Efficient long sequence modeling for time-domain single-channel speech separation. In ICASSP 2020 (pp. 46–50). IEEE.

Macune, D. T., Flanagan, W. D., Choi, E., & Marcellus, E. (2006). A compact compensated resistivity tool for logging while drilling. In SPE/IADC Drilling Conference and Exhibition. SPE.

Mikołajewska, E., Mikołajewski, D., Mikołajczyk, T., & Paczkowski, T. (2025). Generative AI in AI-based digital twins for fault diagnosis for predictive maintenance in Industry 4.0/5.0. Applied Sciences, 15(6), 3166.

Mora-de-León, L. P., Solís-Martín, D., Galán-Páez, J., & Borrego-Díaz, J. (2025). Text-conditioned diffusion-based synthetic data generation for turbine engine sensor analysis and RUL estimation. Machines, 13(5), 374.

Motahhari, H. R., Hareland, G., & James, J. A. (2010). Improved drilling efficiency technique using integrated PDM and PDC bit parameters. Journal of Canadian Petroleum Technology, 49(10), 45–52.

Okoh, C., Roy, R., Mehnen, J., & Redding, L. (2014). Overview of remaining useful life prediction techniques in through-life engineering services. Procedia CIRP, 16, 158–163.

Phelan, Z., Xing, P., Panja, P., Moore, J., & McLennan, J. (2022). Prediction of formation properties based on drilling data of wells at Utah FORGE site using machine learning. In ARMA US Rock Mechanics/Geomechanics Symposium. ARMA.

Rickard, W. M., McLennan, J., Islam, N., & Rivas, E. (2019). Mechanical specific energy analysis of the FORGE Utah well. In Proceedings of the 44th Workshop on Geothermal Reservoir Engineering. Stanford University.

Rykov, Y. (2024). Drilling and logging equipment reliability in a downhole vibration environment. Journal of Petroleum Technology.

Saad, O. M., & Chen, Y. (2020). Deep denoising autoencoder for seismic random noise attenuation. Geophysics, 85(4), V367–V376.

Salinas, D., Flunkert, V., Gasthaus, J., & Januschowski, T. (2020). DeepAR: Probabilistic forecasting with autoregressive recurrent networks. International Journal of Forecasting, 36(3), 1181–1191.

Shah, A. B., Wen, Y., Chen, J., Wu, X., & Fu, X. (2024). A deep learning approach to predicting remaining useful life for downhole drilling sensors using synthetic data generation. In Unconventional Resources Technology Conference. URTeC.

Srivastava, N., Hinton, G., Krizhevsky, A., Sutskever, I., & Salakhutdinov, R. (2014). Dropout: A simple way to prevent neural networks from overfitting. Journal of Machine Learning Research, 15, 1929–1958.

Sutharssan, T., Stoyanov, S., Bailey, C., & Yin, C. (2015). Prognostic and health management for engineering systems: A review of the data-driven approach and algorithms. Journal of Engineering, 2015(7), 215–222.

Urdaneta, C., Jeong, C., Wu, X., & Chen, J. (2024a). Deep learning method for improving rate of penetration prediction in drilling. SPE Journal.

Urdaneta, C., Jeong, C., Wu, X., & Chen, J. (2024b). Predicting coiled-tubing drilling dynamics using transformers. In Unconventional Resources Technology Conference. URTeC.

Urdaneta, C., Jeong, C., & Zheng, A. (2024). An artificial intelligence model to synthesize measurements while drilling sensors for coiled tubing drilling. In SPE/ICoTA Well Intervention Conference and Exhibition. SPE.

Urdaneta, C., Jeong, C., Wu, X., & Chen, J. (2025). Detecting resistivity tool malfunctions in coiled tubing drilling using generative artificial intelligence. In Unconventional Resources Technology Conference. URTeC.

Urdaneta, C., Jarrot, A., Wang, S., Wu, X., & Chen, J. (2023). Deep learning methods for improving electromagnetic telemetry signal-to-noise ratio. SPE Journal, 28(4), 1676–1689.

Vaswani, A., Shazeer, N., Parmar, N., Uszkoreit, J., Jones, L., Gomez, A. N., Kaiser, Ł., & Polosukhin, I. (2017). Attention is all you need. Advances in Neural Information Processing Systems.

Wang, J., Sajeev, S. K., Ozbayoglu, E., Baldino, S., Liu, Y., & Jing, H. (2023). Reducing NPT using a novel approach to real-time drilling data analysis. In SPE Annual Technical Conference and Exhibition. SPE.

Wang, H., Wang, Y., Ding, J., & Wang, J. (2024). Generative adversarial networks for synthetic bearing fault data augmentation: A review. IEEE Access, 12, 28456–28471.

Weiss, K., Khoshgoftaar, T. M., & Wang, D. (2016). A survey of transfer learning. Journal of Big Data, 3(1), 1–40.

Wen, R., Torkkola, K., Narayanaswamy, B., & Madeka, D. (2017). A multi-horizon quantile recurrent forecaster. arXiv preprint arXiv:1711.11053.

Wu, H., Xu, J., Wang, J., & Long, M. (2021). Autoformer: Decomposition transformers with autocorrelation for long-term series forecasting. Advances in Neural Information Processing Systems, 34, 22419–22430.

Yang, L., Wang, S., Chen, X., Saad, O. M., Chen, W., Oboue, Y. A. S. I., & Chen, Y. (2021). Unsupervised 3-D random noise attenuation using deep skip autoencoder. IEEE Transactions on Geoscience and Remote Sensing, 60, 1–16.

Zhou, H., Zhang, S., Peng, J., Zhang, S., Li, J., Xiong, H., & Zhang, W. (2021). Informer: Beyond efficient transformer for long sequence time-series forecasting. In Proceedings of the AAAI Conference on Artificial Intelligence (Vol. 35, pp. 11106–11115).

Zhuang, F., Qi, Z., Duan, K., Xi, D., Zhu, Y., Zhu, H., Xiong, H., & He, Q. (2021). A comprehensive survey on transfer learning. Proceedings of the IEEE, 109(1), 43–76.

Synthetic Data-Driven Hybrids for Deep Learning Drilling Models

8

8.1 Motivation and Contribution Overview

Previous chapter demonstrated that synthetic data generation enables accurate RUL prediction on the FORGE geothermal dataset, achieving 14.80% mean absolute percentage error. However, models trained on single-site data struggle when deployed to wells with different operational characteristics. When FORGE-trained models are tested on the Volve offshore dataset, prediction errors exceed 20% due to shifts in temperature distributions, pressure regimes, and drilling practices.

This chapter addresses the generalization challenge by harmonizing FORGE and Volve datasets through extended synthetic augmentation and introducing a hybrid BiGRU-BiLSTM architecture. The bidirectional design captures both early warning signals and late-stage nonlinear degradation patterns that unidirectional models miss. Experimental results show the hybrid achieves 14.92% error on combined data, enabling reliable cross-site deployment with minimal adaptation overhead.

8.2 Background and Related Work

Operational fields differ in sensor types, unit conventions, and drilling regimes, making generalization a central challenge for scalable predictive maintenance. Shah et al. (2024) demonstrated initial success with LSTM-based RUL prediction on single-site geothermal data but noted that performance degraded when models encountered unseen formation types or equipment configurations. This observation aligns with broader findings in transfer learning for industrial prognostics, where domain shift i.e. differences in sensor noise characteristics, operational envelope boundaries, and degradation mechanisms limits the applicability of site-specific models (Zhuang et al. 2021; Weiss et al. 2016).

8.2.1 Synthetic Data Augmentation Strategies

The literature on synthetic augmentation for time-series prognostics includes several complementary approaches. Generative adversarial networks (GANs) and variational autoencoders (VAEs) have been applied to bearing vibration data (Wang et al. 2024) and turbine sensor streams (Mora-de León et al. 2025), demonstrating improved model robustness to rare failure modes. However, these generative models require substantial training data and often produce artifacts that lack physical interpretability. Diffusion models represent a recent advancement, offering higher-fidelity sample generation but at significant computational cost (Kong et al. 2023).

C. Urdaneta et al., *Deep Learning Applications for Drilling Performance: Forecasting, Diagnostics, Telemetry, and Predictive Maintenance*, Advances in Oil and Gas Exploration & Production,
https://doi.org/10.1007/978-3-032-26648-4_8

In contrast, parametric augmentation, where synthetic samples are generated through physics-informed transformations of real measurements remains prevalent in drilling and subsurface applications. This approach offers three key advantages: (1) generated samples inherit domain-grounded structure, (2) augmentation parameters can be calibrated using engineering expertise, and (3) computational overhead is minimal compared to training generative models. Prior work by Shah et al. (2024) established temperature-based degradation as a viable parametric framework for downhole sensors, but their methodology was limited to single-site deployment.

8.2.2 Dataset Harmonization

Public datasets rarely contain full run to failure histories, yet models must see late-life, non-linear behavior to be useful in the field, so we use synthetic generation to fill this gap by producing realistic failure trajectories that remain grounded in domain physics while preserving the statistical character of observed telemetry. We begin by harmonizing and cleaning source traces, standardizing units, mapping sensors to a common schema, and resampling time series to a consistent cadence. We apply per-site normalization to preserve local behavior while enabling cross-site training, and we partition data by equipment for validation so that performance reflects true generalization rather than temporal leakage. From this harmonized base we synthesize trajectories by sampling lifespan parameters from empirically motivated distributions, applying multiplicative offsets and slow bias drifts to temperature channels, and introducing occasional step changes, abrupt drops to near zero and transient spikes at calibrated rates, all under fixed random seeds to ensure reproducibility. Our objective is to span plausible degradation and fault modes without creating unrealistic artifacts. Finally, we validate the synthetic ensembles with a battery of statistical and temporal checks, for example Kolmogorov–Smirnov tests on marginals, covariance matrix comparisons to assess multivariate structure, and dynamic time warping to verify temporal alignment, which ensures the synthetic data preserve the key properties of real telemetry while expanding coverage of rare late-life behaviors.

8.2.3 Temporal Anomaly Injection and Drift Modeling

To mimic realistic failure precursors, we inject temporal anomalies and slow drifts in ways that respect observed temporal scales. We model slow degradation as a low-frequency bias term, for example a small-step random walk or a low-order autoregressive component, and we superimpose occasional step changes to emulate installation issues or abrupt sensor degradation. For intermittent faults we introduce short-duration zero or near-zero readings and transient spikes according to a calibrated Poisson process whose event rate increases with cumulative stress indicators such as cumulative temperature exposure. When injecting anomalies, we ensure amplitudes and durations remain plausible and we avoid patterns that are never seen in the field, because such artifacts can mislead models. For reproducibility we record random seeds so that each synthetic dataset and each individual synthetic event can be regenerated and inspected.

8.2.4 Feature Enhancement and Visualization

To ensure the quality and interpretability of our dataset we apply a feature enhancement step after cleaning and harmonization, removing records with irrecoverable nulls, flagging extreme outliers for inspection, and then adding lifecycle labeling. specifically, we introduce a life cycle column that increments by one for each row so every record has a unique temporal index. To validate the synthetic ensembles and support operator trust, we produce well log style visualizations that overlay key channels such as raw temperature, pump pressure and a scaled probability-of-zero indicator for representative sensors. These are presented in Fig. 8.1.

From the well logs it can be observed that the sensors operate within similar thermal bands while showing gradual temperature shifts and episodic excursions that become more pronounced toward late life. The pump pressure panel reveals repeating operational cycles and occasional transient peaks, while the probability-of-zero trace highlights intermittent dropouts or injected faults, and its co-location with temperature and pressure features supports the likelihood of the synthetic anomalies and their relevance for RUL prediction.

8.2.5 Validation Metrics and Acceptance Criteria

We establish clear, quantitative acceptance criteria to decide when a synthetic ensemble is fit for training prognostic models. At the univariate level we apply Kolmogorov–Smirnov tests on key channels, and at multivariate level we compare covariance structures using matrix norms. For temporal fidelity we use dynamic time warping distances and inspect autocorrelation functions, and when models depend on rare anomalies, we also evaluate extreme value behavior and event inter-arrival statistics. In addition to these statistical checks, we validate downstream utility by training a fast proxy model on the synthetic data and

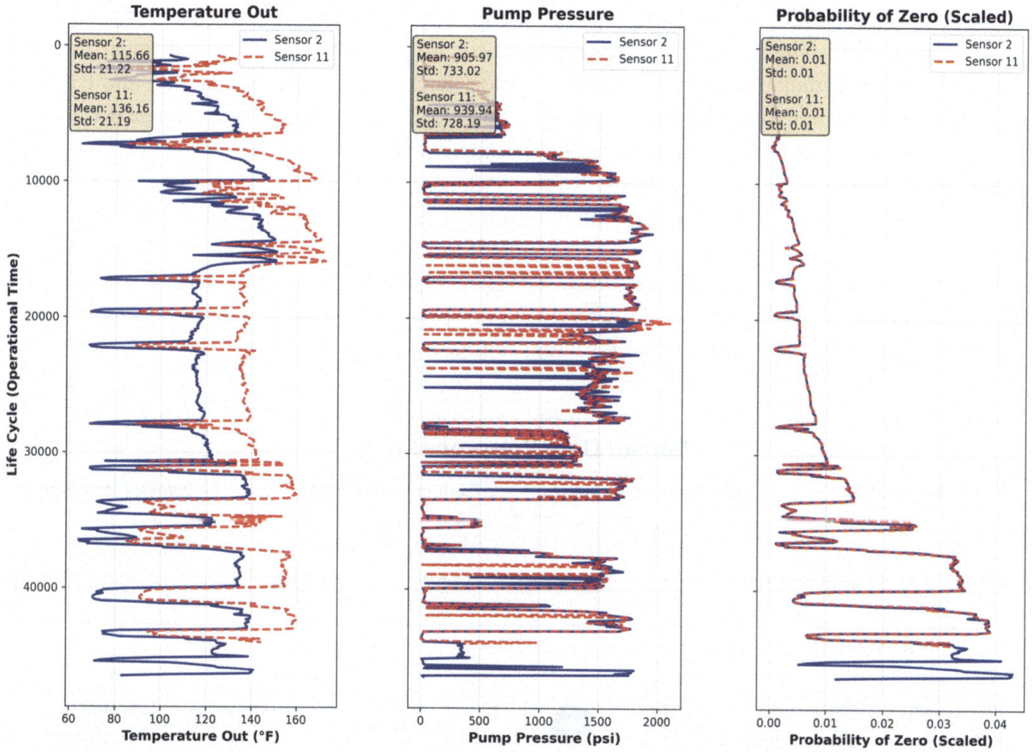

Fig. 8.1 Well log style comparison of key sensor features, showing temporal evolution over operational life cycle

measuring performance on a small set of held-out real trajectories; parity or improvement in test metrics indicates practical usefulness. We document threshold values for each metric that represent acceptable fidelity and iterate synthesis parameters until ensembles consistently meet those thresholds.

8.3 Hybrid Model Design and Training Methodology

Gate recurrent unit (GRU) and long short-term memory (LSTM) units offer complementary strengths that suit time series prognostics, and we design our hybrid to exploit those strengths. GRUs are parameter efficient and train faster, which makes them useful for encoding local and medium-range temporal patterns with lower computational cost, while LSTM cells excel at preserving long-range dependencies and filtering noisy signals through gated cell states, which is important for modeling slow cumulative degradation. Bidirectionality further improves the model's ability to detect both early and late-stage degradation by letting the network consider context from past and future timesteps during representation learning.

We architect the model pipeline to follow a clear, reproducible flow from raw telemetry to remaining useful life prediction, beginning with raw sensor data ingestion where telemetry, operational annotations and metadata are collected and stored in a canonical format, then applying systematic preprocessing steps such as unit standardization, gap handling, outlier mitigation and time resampling so downstream computations operate on clean, consistent signals and spurious variability is reduced, and finally computing an engineered feature set that blends physics-informed indicators with statistical summaries so that every timestep carries domain knowledge, alongside multi-scale context provided by rolling means, variances and first differences. The workflow is shown in Fig. 8.2.

We place bidirectional GRU layers at the model input to quickly build a compact, context rich representation of recent and near past behavior, and we follow them with bidirectional LSTM layers that consolidate that representation into a stable long-range memory. This ordering reduces parameter cost in the front end while allowing the deeper LSTM stack to focus on capturing complex non-linear degradation dynamics. Combining BiGRU and BiLSTM units lets us obtain a compact, context rich front end and then consolidate those encodings into a deeper, long memory representation that is effective for remaining useful life forecasting. While bidirectionality enhances both units by allowing the network to consider context from both past and future timesteps during representation learning.

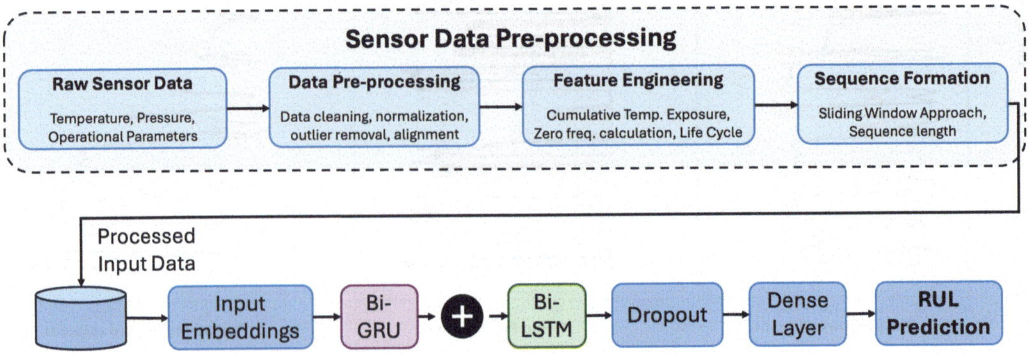

Fig. 8.2 Workflow of the synthetic data-driven hybrid deep learning model, implementing a Bi-directional GRU-LSTM architecture on a combined dataset of FORGE and Volve datasets

8.3.1 Mathematical Formulation and Architecture Detail

We pose remaining useful life (RUL) prediction as a supervised sequence regression problem. Given a length-k input $X_t = x_{t-k+1}, x_{t-k+2}, \ldots, x_t$ represent a sequence of k consecutive multivariate observations up to time t, where each $x_i \in \mathbb{R}^d$ is a d dimensional feature vector. Our objective is to learn a function f_θ :

$$\widehat{y}_t = f_\theta X_t \tag{8.1}$$

where $\widehat{y}_t$ representes the predicted RUL at time t. The learning objective is to find the optimal parameters θ that minimize the expected prediction error over the distribution of possible input sequences and their corresponding true RUL values:

$$\theta^* = \underset{\theta}{\operatorname{argmin}} \mathbb{E}_{\{X,y\} \sim \mathfrak{D}}[L(f_\theta(X), y)] \tag{8.2}$$

where $\mathfrak{D}$ represents the data distribution, L is a loss function measuring the discrepancy between predicted and true RUL values and y is the ground truth RUL value. For this specific problem, we employ Mean Squared Error (MSE) as the loss function:

$$L(f_\theta(\mathrm{X}, y) = (f_\theta(\mathrm{X}) - y)^2 \tag{8.3}$$

A key challenge in RUL prediction is capturing both short-term sensor variations and long-term degradation trends across diverse, non-stationary environments. We therefore implement f by first encoding the input sequence with bidirectional GRU layers. For an input sequence of $\mathrm{X} = \{x_1, x_2, \ldots, x_k\}$ the BiGRU layer computes both forward ($\overrightarrow{h}_t$) and backward ($\overleftarrow{h}_t$) hidden states:

$$\overrightarrow{h}_t = GRUf\left(x_t, \overrightarrow{h}_{t-1}\right) \tag{8.4}$$

$$\overleftarrow{h}_t = GRUf(x_t, \overleftarrow{h}_{t+1}) \tag{8.5}$$

where the GRU operation at each time step involves update z_t and reset r_t gates, along with candidate hidden state $\tilde{h}_t$. The bidirectional outputs are then concatenated to form a comprehensive representation:

$$h_t^{BiGRU} = [\overrightarrow{h}_{t;} \overleftarrow{h_t}] \tag{8.6}$$

This BiGRU output sequence is then processed by a Bidirectional LSTM layer, which maintains separate memory cell states c_t and hidden states h_t with input i_t, forget f_t, output o_t, and cell candidate $\tilde{c}_t$ gates. Similar to the BiGRU layer, the BiLSTM processes the sequence and the outputs are concatenated:

$$h_t^{BiLSTM} = [\overrightarrow{h}_t^{LSTM}; \overleftarrow{h}^{LSTM}] \tag{8.7}$$

The final BiLSTM hidden state passes through dropout, a ReLU-activated dense layer, and a linear output to predict RUL. This structure captures both short-term signals and long-term trends, with bidirectional layers leveraging past and future context. The model balances expressiveness, efficiency, and overfitting control with a manageable parameter count.

8.3.2 Training Regimen and Implementation Details

We implement a training regimen that preserves temporal ordering and prevents data leakage across equipment. We combine the FORGE and Volve datasets, applying per-site normalization followed by global rescaling. To ensure robust generalization, we employ k-fold cross-validation (k = 5) stratified by unique equipment IDs, ensuring that each fold contains hold-out equipment entirely unseen during training. We train with mean squared error (MSE) as the objective and monitor normalized RMSE and MAPE on validation folds. Learning rate schedules, early stopping and gradient clipping are applied to promote stable convergence. When evaluating cross-field generalization we report both aggregated metrics and per-site breakdowns so that transportability issues become evident.

We implement all models using standard Python deep learning frameworks for data loading, augmentation, and training. Hyperparameter optimization is performed through grid searches over the following ranges: hidden layer sizes (64–256), recurrent layers (2–4), dropout rates (0.1–0.4), batch sizes (32–128), and initial learning rates (1×10^{-4}–1×10^{-3}). Final model parameters are selected based on optimal validation performance.

8.4 Experimental Evaluation and Results

8.4.1 Experimental Setup and Data Splits

We evaluate the hybrid BiGRU–BiLSTM model on the harmonized FORGE and Volve corpus described earlier. After harmonization and augmentation, the combined dataset contains hundreds of thousands of timesteps spanning diverse operating regimes. For dataset partitioning we used a stratified temporal split (70–15–15) to partition the combined FORGE and Volve dataset, preserving chronological order and degradation phase balance. We also apply per-site normalization prior to global rescaling and use equipment-wise partitioning to avoid leakage. All experiments were conducted using PyTorch on a computing infrastructure equipped with NVIDIA GPU featuring 40 GB VRAM.

8.4.2 Baselines and Evaluation Metrics

We evaluate the performance of our model using a comprehensive experimental framework designed to assess both prediction accuracy and generalization capabilities across diverse operational conditions. We conducted comparative analysis against several baseline models, including traditional machine learning methods (Random Forest, Gradient Boosting) and standard deep learning architectures (vanilla LSTM, CNN-LSTM, GRU, standalone BiGRU). To ensure a fair comparison, all models were trained and evaluated using identical training, validation, and testing partitions, with consistent performance metrics across models. Initially, we assessed the models' ability to predict RUL values (GT) derived from the synthetic dataset. Performance was then further scrutinized using key metrics such as RMSE, MAPE, and R2 Score.

Results in Fig. 8.3 presents predicted RUL for selected sensors from both the FORGE and Volve datasets. The visualizations reveal that the model accurately captures both the overall trend and the specific degradation patterns unique to each sensor. The model, compared to baselines, not only captures the downward trend but also adapts to local fluctuations with minimal deviation from the true RUL trajectory. This visual consistency across diverse sensors and operational domains further supports the model's generalizability and reliability in real-world predictive maintenance applications.

8.5 Conclusion

We presented a practical pipeline that combines physics-aligned synthetic run-to-failure generation with a hybrid BiGRU–BiLSTM architecture to improve RUL prediction for downhole drilling assets. Our method harmonizes multi-site telemetry, synthesizes realistic drift and anomaly patterns, and consolidates efficient BiGRU front ends with memory-rich BiLSTM stacks to model both local variability and long-term degradation. Experimental evaluation on the combined FORGE and Volve corpus shows that synthetic-informed hybrids improve late-life forecasting accuracy relative to single-unit recurrent models and tree-based baselines, and that uncertainty-aware variants provide calibrated predictive intervals that are actionable for maintenance planning.

Operationally, we find that pretraining on combined real and synthetic ensembles accelerates on-site adaptation via modest fine-tuning, and that standard compression strategies make

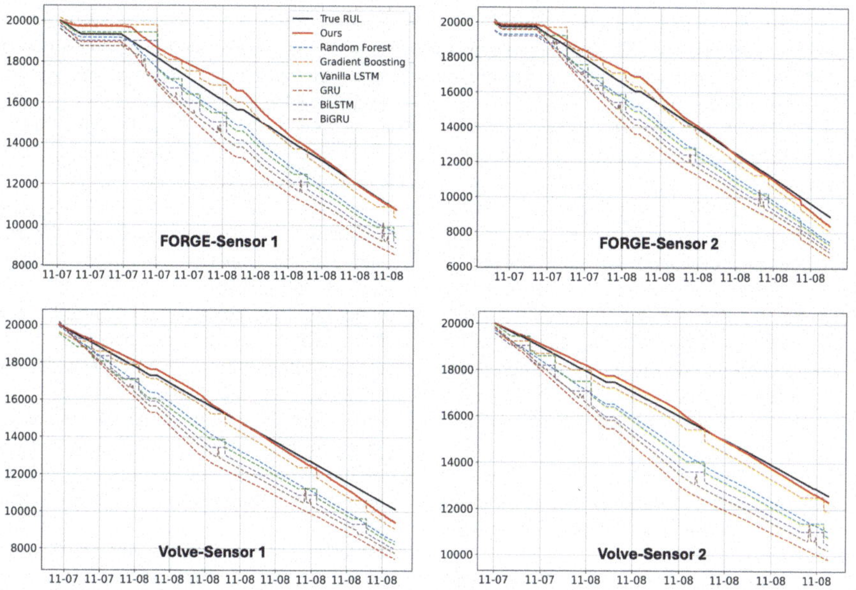

Fig. 8.3 Comparison of RUL predictions across models on the FORGE and Volve datasets. Our proposed model (red) exhibits the highest alignment with True RUL (black), surpassing all baseline models especially in non-linear instances

the hybrid amenable to edge deployment. For production use we emphasize rigorous validation, clear governance and traceability of synthetic datasets, and a staged rollout that begins with human-in-the-loop monitoring.

Future work should extend synthesis to vibration, acoustic and chemical modalities, automate synthesis calibration via meta-learning, and embed decision-aware objectives so models optimize directly for operational value under cost and availability constraints.

References

Abbaszadeh Shahri, A., Shan, C., Larsson, S., & Johansson, F. (2024). Normalizing large scale sensor based MWD data: An automated method toward a unified database. Sensors, 24(4), 1209.

Al-Fakih, A., Koeshidayatullah, A., Mukerji, T., & Kaka, S. I. (2024). Enhanced anomaly detection in well log data through the application of ensemble GANs. arXiv preprint arXiv:2411.19875.

Alsaihati, A., & Alotaibi, B. (2022). Determining severity of lateral and torsional downhole vibrations while drilling surface holes using three machine learning techniques. SPE Journal, 27(3), 1493–1503.

Altindal, M. C., Nivlet, P., Tabib, M., Rasheed, A., Kristiansen, T. G., & Khosravanian, R. (2024). Anomaly detection in multivariate time series of drilling data. Geoenergy Science and Engineering, 237, 212778.

Aoun, M. A. B., & Madarász, T. (2022). Applying machine learning to predict the rate of penetration for geothermal drilling located in the Utah FORGE site. Energies, 15(12), 4288.

Atashnezhad, A., Akhtarmanesh, S., Hareland, G., & Al Dushaishi, M. (2021). Developing a drilling optimization system for improved overall rate of penetration in geothermal wells. In ARMA US Rock Mechanics/Geomechanics Symposium. ARMA.

Bergstra, J., & Bengio, Y. (2012). Random search for hyper-parameter optimization. Journal of Machine Learning Research, 13, 281–305.

Biggio, L., Bendinelli, T., Kulkarni, C., & Fink, O. (2023). Ageing-aware battery discharge prediction with deep learning. Applied Energy, 346, 121229.

Carpenter, C. (2015). Review of downhole wireless communication techniques. Journal of Petroleum Technology, 67(3), 120–122.

Champion, B. P. (2016). Wireless reservoir monitoring reducing reservoir uncertainty during Barents Sea appraisal: A case history for Norvarg. In SPE Norway Subsurface Conference. SPE.

Ding, Z., Chen, Y., Wang, H., Wang, X., Zhang, W., & Zhang, Y. (2025). TIformer: A transformer-based framework for time-series forecasting with missing data. In Australasian Database Conference (pp. 71–84). Springer.

Dinh, N. M. (2020). Early kick detection using data-driven Bayesian network: Model development and experimental testing (Doctoral dissertation, Memorial University of Newfoundland).

Du, S., Huang, C., Ma, X., & Fan, H. (2024). A review of data-driven intelligent monitoring for geological drilling processes. Processes, 12(11), 2478.

Gidh, Y., Purwanto, A., & Smith Bits. (2012). Artificial neural network drilling parameter optimization system improves ROP by predicting and managing bit wear. In SPE Intelligent Energy International Conference and Exhibition. SPE.

Girase, H., Agarwal, N., Choi, C., & Mangalam, K. (2023). Latency matters: Real-time action forecasting transformer. In Proceedings of the IEEE/CVF Conference on Computer Vision and Pattern Recognition (pp. 18759–18769).

Gwynn, M., Allis, R., Hardwick, C., Jones, C., Nielsen, P., & Hurlbut, W. (2018). Rock properties of FORGE Well 58–32, Milford, Utah. Transactions, Geothermal Resources Council.

Habashy, T., & Anderson, B. I. (1991). Reconciling differences in depth of investigation between 2-MHz phase shift and attenuation resistivity measurements. In SPWLA Annual Logging Symposium. SPWLA.

Hegde, C., Soares, C., & Gray, K. (2018). Rate of penetration modeling using hybrid models: Deterministic and machine learning. In Unconventional Resources Technology Conference (pp. 3220–3238).

Hermawan Manuaba, I. B. G., Aljishi, M., Van Steene, M., & Dolan, J. (2024). Logging-while-drilling laterolog vs. electromagnetic propagation measurements: Which is telling the true resistivity? SPE Journal, 29(8), 4000–4013.

Hyndman, R. J. (2018). Forecasting: Principles and practice. OTexts.

Iglesias, G., Talavera, E., González-Prieto, Á., Mozo, A., & Gómez-Canaval, S. (2023). Data augmentation techniques in time series domain: A survey and taxonomy. Neural Computing and Applications, 35(14), 10123–10145.

Islam, N., Vijapurapu, R., Jones, M., McLennan, J., Moore, J., Rickard, W., Balamir, O., & Vagnetti, R. (2018). Application of mechanical specific energy and at-the-bit measurements for geothermal drilling applications. In ARMA US Rock Mechanics/Geomechanics Symposium. ARMA.

Jannin, G., Chen, J., DePavia, L. E., Sun, L., & Schwartz, M. (2017). Deep electrode: A game-changing technology for electromagnetic telemetry. In SEG International Exposition and Annual Meeting. SEG.

Keman, L. (2016). Adaptive noise cancellation for electromagnetic-while-drilling system. In International Conference on Information Science and Control Engineering (pp. 1253–1256). IEEE.

Kingma, D. P., & Welling, M. (2019). An introduction to variational autoencoders. Foundations and Trends in Machine Learning, 12(4), 307–392.

Koeshidayatullah, A., Al-Fakih, A., & Kaka, S. I. (2024). Leveraging time-series foundation model for subsurface well logs prediction and anomaly detection. arXiv preprint arXiv:2412.05681.

Kong, Z., Xu, Y., Zhang, Z., & Luo, B. (2023). Diffusion models for time series applications: A survey. arXiv preprint arXiv:2305.00624.

Luo, Y., Chen, Z., & Yoshioka, T. (2020). Dual-path RNN: Efficient long sequence modeling for time-domain single-channel speech separation. In ICASSP 2020 (pp. 46–50). IEEE.

Macune, D. T., Flanagan, W. D., Choi, E., & Marcellus, E. (2006). A compact compensated resistivity tool for logging while drilling. In SPE/IADC Drilling Conference and Exhibition. SPE.

Mikołajewska, E., Mikołajewski, D., Mikołajczyk, T., & Paczkowski, T. (2025). Generative AI in AI-based digital twins for fault diagnosis for predictive maintenance in Industry 4.0/5.0. Applied Sciences, 15(6), 3166.

Mora-de-León, L. P., Solís-Martín, D., Galán-Páez, J., & Borrego-Díaz, J. (2025). Text-conditioned diffusion-based synthetic data generation for turbine engine sensor analysis and RUL estimation. Machines, 13(5), 374.

Motahhari, H. R., Hareland, G., & James, J. A. (2010). Improved drilling efficiency technique using integrated PDM and PDC bit parameters. Journal of Canadian Petroleum Technology, 49(10), 45–52.

Okoh, C., Roy, R., Mehnen, J., & Redding, L. (2014). Overview of remaining useful life prediction techniques in through-life engineering services. Procedia CIRP, 16, 158–163.

Phelan, Z., Xing, P., Panja, P., Moore, J., & McLennan, J. (2022). Prediction of formation properties based on drilling data of wells at Utah FORGE site using machine learning. In ARMA US Rock Mechanics/Geomechanics Symposium. ARMA.

Rickard, W. M., McLennan, J., Islam, N., & Rivas, E. (2019). Mechanical specific energy analysis of the FORGE Utah well. In Proceedings of the 44th Workshop on Geothermal Reservoir Engineering. Stanford University.

Rykov, Y. (2024). Drilling and logging equipment reliability in a downhole vibration environment. Journal of Petroleum Technology.

Saad, O. M., & Chen, Y. (2020). Deep denoising autoencoder for seismic random noise attenuation. Geophysics, 85(4), V367–V376.

Salinas, D., Flunkert, V., Gasthaus, J., & Januschowski, T. (2020). DeepAR: Probabilistic forecasting with autoregressive recurrent networks. International Journal of Forecasting, 36(3), 1181–1191.

Shah, A. B., Wen, Y., Chen, J., Wu, X., & Fu, X. (2024). A deep learning approach to predicting remaining useful life for downhole drilling sensors using synthetic data generation. In Unconventional Resources Technology Conference. URTeC.

Srivastava, N., Hinton, G., Krizhevsky, A., Sutskever, I., & Salakhutdinov, R. (2014). Dropout: A simple way to prevent neural networks from overfitting. Journal of Machine Learning Research, 15, 1929–1958.

Sutharssan, T., Stoyanov, S., Bailey, C., & Yin, C. (2015). Prognostic and health management for engineering systems: A review of the data-driven approach and algorithms. Journal of Engineering, 2015(7), 215–222.

Urdaneta, C., Jeong, C., Wu, X., & Chen, J. (2024a). Deep learning method for improving rate of penetration prediction in drilling. SPE Journal.

Urdaneta, C., Jeong, C., Wu, X., & Chen, J. (2024b). Predicting coiled-tubing drilling dynamics using transformers. In Unconventional Resources Technology Conference. URTeC.

Urdaneta, C., Jeong, C., & Zheng, A. (2024). An artificial intelligence model to synthesize measurements while drilling sensors for coiled tubing drilling. In SPE/ICoTA Well Intervention Conference and Exhibition. SPE.

Urdaneta, C., Jeong, C., Wu, X., & Chen, J. (2025). Detecting resistivity tool malfunctions in coiled tubing drilling using generative artificial intelligence. In Unconventional Resources Technology Conference. URTeC.

Urdaneta, C., Jarrot, A., Wang, S., Wu, X., & Chen, J. (2023). Deep learning methods for improving electromagnetic telemetry signal-to-noise ratio. SPE Journal, 28(4), 1676–1689.

Vaswani, A., Shazeer, N., Parmar, N., Uszkoreit, J., Jones, L., Gomez, A. N., Kaiser, Ł., & Polosukhin, I. (2017). Attention is all you need. Advances in Neural Information Processing Systems.

Wang, J., Sajeev, S. K., Ozbayoglu, E., Baldino, S., Liu, Y., & Jing, H. (2023). Reducing NPT using a novel approach to real-time drilling data analysis. In SPE Annual Technical Conference and Exhibition. SPE.

Wang, H., Wang, Y., Ding, J., & Wang, J. (2024). Generative adversarial networks for synthetic bearing fault data augmentation: A review. IEEE Access, 12, 28456–28471.

Weiss, K., Khoshgoftaar, T. M., & Wang, D. (2016). A survey of transfer learning. Journal of Big Data, 3(1), 1–40.

Wen, R., Torkkola, K., Narayanaswamy, B., & Madeka, D. (2017). A multi-horizon quantile recurrent forecaster. arXiv preprint arXiv:1711.11053.

Wu, H., Xu, J., Wang, J., & Long, M. (2021). Autoformer: Decomposition transformers with auto-correlation for long-term series forecasting. Advances in Neural Information Processing Systems, 34, 22419–22430.

Yang, L., Wang, S., Chen, X., Saad, O. M., Chen, W., Oboue, Y. A. S. I., & Chen, Y. (2021). Unsupervised 3-D random noise attenuation using deep skip autoencoder. IEEE Transactions on Geoscience and Remote Sensing, 60, 1–16.

Zhou, H., Zhang, S., Peng, J., Zhang, S., Li, J., Xiong, H., & Zhang, W. (2021). Informer: Beyond efficient transformer for long sequence time-series forecasting. In Proceedings of the AAAI Conference on Artificial Intelligence (Vol. 35, pp. 11106–11115).

Zhuang, F., Qi, Z., Duan, K., Xi, D., Zhu, Y., Zhu, H., Xiong, H., & He, Q. (2021). A comprehensive survey on transfer learning. Proceedings of the IEEE, 109(1), 43–76.

9 Challenges, Limitations, and Future Directions

This chapter documents limitations that are specific to the datasets, modeling choices, and evaluation setups used in Chaps. 3 through 8. The intent is to make clear what the reported results do and do not establish, and which constraints would need to be addressed before transferring these workflows to new rigs, fields, or drilling programs. The limitations are grouped into three categories: dataset limitations, method and model limitations, and deployment constraints that follow from the preceding two categories. While some limitations are common across drilling applications of data driven modeling, the emphasis here is on the concrete failure modes and boundary conditions implied by the specific case studies in this volume.

9.1 Data Quality, Volume, and Labeling Constraints

The datasets used in Chaps. 3 through 8 differ in sampling rate, sensor type, and operational context, and these differences influence what can be concluded from each case study. In forecasting chapters, the target variables reflect both formation response and operational policy. In diagnostics and denoising chapters, the observed signals include tool specific artifacts that may not be stable across hardware generations. In predictive maintenance chapters, labels are commonly incomplete or delayed relative to the underlying degradation process. These properties introduce identifiable limitations that should be treated as part of the results (Du et al. 2024).

9.1.1 Chapter 3 Dataset Limitations for Geothermal Rate of Penetration Forecasting

The geothermal forecasting results are conditional on the coverage of the Utah FORGE time series used for model development. The data represent a specific site, well configuration, and drilling program, and the trained model therefore learns statistical regularities associated with that operating envelope (Aoun and Madarász 2022). Even when a multi-section evaluation is performed within the dataset, the model is still exposed to correlated conditions such as consistent sensor suites, consistent data processing conventions, and related operational practices. As a result, generalization claims should be framed as generalization within the evaluated wells and intervals, rather than generalization to geothermal drilling broadly. Additional uncertainty arises from the practical realities of geothermal environments, where elevated temperature, vibration, and tool wear can drive sensor drift and intermittent measurement artifacts. These effects can shift the data distribution over time, including within a single well, which

C. Urdaneta et al., *Deep Learning Applications for Drilling Performance: Forecasting, Diagnostics, Telemetry, and Predictive Maintenance*, Advances in Oil and Gas Exploration & Production,
https://doi.org/10.1007/978-3-032-26648-4_9

limits the stability of a single model trained on historical intervals.

Future work that aims to strengthen external validity should prioritize evaluation across distinct wells, bit runs, and operational regimes that are known to produce distribution shift. Where possible, model performance should be reported separately for intervals with stable sensor behavior and intervals flagged for drift or dropout, since these categories reflect different inference conditions.

9.1.2 Chapter 4 Dataset Limitations for Coiled Tubing Drilling Dynamics Forecasting

The coiled tubing drilling forecasting results are constrained by two related issues: heterogeneity and label definition. First, coiled tubing datasets tend to mix transient events, steady drilling intervals, and operational transitions within short time windows. This heterogeneity is part of the real operational context, but it also means that model accuracy depends strongly on the proportion of each regime present in training and testing windows. Second, shock metrics and related indicators can be sensitive to sensor placement, sampling frequency, and processing pipelines (Li et al. 2021). If shock measurements are aggregated or filtered differently across runs, then the learned mapping can partially reflect instrumentation conventions rather than downhole physics.

A direct implication is that the reported accuracy should be interpreted as evidence that the transformer architecture can learn useful temporal structure in the available coiled tubing series, not as a guarantee that the same trained weights will transfer to new coiled tubing assemblies or new sensor definitions. A practical mitigation, aligned with the workflow presented in this book, is to treat the model as a component within an adaptive monitoring system where periodic recalibration is expected when tool strings or acquisition settings change.

9.1.3 Chapter 5 Dataset Limitations for Resistivity Tool Diagnostics

The resistivity diagnostics study relies on a small set of measurement channels and a limited number of wells, with labeled malfunction intervals that represent a subset of the failure modes observed in field operations. The preprocessing choices also impose explicit constraints. For example, the approach avoids gap filling and excludes discontinuous intervals from training, which reduces the risk of injecting artificial patterns but also limits the model's exposure to realistic telemetry dropouts and operational interruptions.

The anomaly scoring behavior is therefore best interpreted as performance under continuous logging conditions similar to the curated training segments, rather than performance across all operational conditions (Altindal et al. 2024).

Additionally, the model is trained to reconstruct nominal behavior from healthy intervals and then flags deviation. This setup depends on the assumption that the healthy training intervals span the normal variability of tool response. If the tool operates in a new lithology, with different attenuation and phase shift characteristics, nominal behavior may shift even without malfunction, and reconstruction error may increase for reasons unrelated to tool health. This creates a known ambiguity between formation driven shift and tool driven degradation (Hermawan Manuaba et al. 2024). Future work that seeks to reduce this ambiguity can prioritize adding contextual channels that capture operational state, and reporting error stratified by formation or operational regime where labels exist.

9.1.4 Chapter 6 Dataset Limitations for Electromagnetic Telemetry Denoising

The telemetry denoising results depend on the representativeness of the field noise environment and the structure of the synthetic mixtures used for supervised training phases.

Chapter 6 supports an internally consistent evaluation for the given acquisition setting, but it does not guarantee transfer to different electromagnetic environments, different stake geometries, or different interference sources. If the dominant noise sources change, a model trained on one mixture family may underperform or produce artifacts that look plausible in time domain but distort spectral content relevant to decoding.

A practical limitation is that denoising quality is not fully captured by pointwise error measures alone if the downstream objective is improved demodulation or bit error rate (Mwachaka et al. 2019). Future work that strengthens the connection to operational value should incorporate task aligned metrics, such as decode success rate or proxy measures of signal integrity, alongside waveform reconstruction scores.

9.1.5 Chapters 7 and 8 Dataset Limitations for Predictive Maintenance and Synthetic Hybrid Modeling

Remaining useful life estimation is fundamentally limited by label quality and by censoring. Many tools are replaced for operational reasons before failure, which creates right censored histories where the true remaining life is unobserved. Even when failure occurs, the onset of degradation may precede the logged event by an unknown interval. These properties can bias supervised learning toward patterns that correlate with maintenance policy rather than the degradation process itself (Okoh et al. 2014).

The synthetic and hybrid modeling approaches introduced in the book provide one path to addressing coverage limitations, but they also introduce a separate constraint because synthetic samples reflect the assumptions embedded in the augmentation or simulation procedure. If those assumptions omit rare but operationally important behaviors, synthetic expansion can increase apparent coverage while still missing the critical tails of the distribution. For that reason, synthetic data should be positioned as a tool for stress testing and robustness, but it is not a complete substitute for diverse field data (Iglesias et al. 2023).

9.2 Model Generalization and Overfitting

Across Chaps. 3 through 6, the core modeling components include transformer based forecasters and autoencoder family models for reconstruction based inference. These model classes can learn complex nonlinear mappings, but they can also overfit to dataset specific artifacts, and they may extrapolate with unwarranted confidence under distribution shift (Polo et al. 2023).

9.2.1 Transformer Forecasting Limitations in Chaps. 3 and 4

The transformer forecasters are designed to leverage long and short range dependencies in multivariate drilling time series. A limitation of this approach is that attention mechanisms can prioritize correlations that are stable in the training set but not causal in the field. In drilling data, operational policy often creates structured correlations, for example repeated parameter schedules or responses to specific alarms. When these policies change, model performance can degrade even if formation response remains similar. This risk is elevated when training and testing windows are drawn from the same program, because policy signatures can be shared across splits.

Another limitation concerns horizon dependent error accumulation. Forecast accuracy typically declines as horizon increases, but the rate of decline can vary by section, by drilling regime, and by sensor stability. If a single horizon is selected for reporting, it may mask meaningful regime specific performance differences. A practical mitigation is to report horizon sensitivity and to define use cases where the forecast is treated as a short range decision support signal rather than a long range plan predictor.

9.2.2 Autoencoder and Variational Autoencoder Limitations in Chaps. 5 and 6

Reconstruction based models provide a natural mechanism for anomaly detection and denoising, but they introduce two interpretability constraints. First, reconstruction error is not specific to a root cause. High reconstruction error can reflect tool malfunction, formation transition, changes in acquisition settings, or sensor drift. Second, reconstruction quality can be high even when subtle but operationally meaningful distortions remain, especially when the model learns to smooth signals.

In Chap. 5, the model focuses on pointwise behavior and therefore may miss slow drift patterns that require temporal context to distinguish from normal variability. In Chap. 6, separation and denoising models can introduce inaccurate structure when the input signal is dominated by noise outside the model's training distribution. This is a critical limitation for any workflow where denoised outputs feed into downstream decoding or control logic.

9.2.3 Uncertainty Representation and Operational Interpretability

When model outputs influence operational decisions, point predictions alone are insufficient for safe use. The forecasting chapters include uncertainty quantification and validation strategy elements, but a limitation remains where uncertainty estimates can be miscalibrated under distribution shift. Calibration that holds within one dataset split can fail when sensors drift or when the drilling regime moves outside training coverage. A practical implication is that uncertainty outputs should be paired with distribution shift indicators, so that rising uncertainty can be interpreted.

9.3 Real-Time Processing and Deployment Constraints

The deployment constraints for the workflows in this book follow directly from the dataset and model limitations above. Even when a model performs well offline, operational use requires that inference be timely, that the data stream be stable, and that failure modes be managed explicitly. There are latency and integration constraints for real time drilling contexts. The key limitation for the chapters in this book is that the case studies are evaluated primarily in offline settings (Nunes et al. 2023).

9.3.1 Data Pipeline Dependencies

Each case study assumes a defined preprocessing pipeline, including scaling choices, windowing, and exclusion rules. In deployment, these steps must be reproduced consistently, and small differences can create large performance changes. For example, changes in normalization statistics, resampling, or dropout handling can shift the effective input distribution. A practical mitigation is to treat the preprocessing pipeline as part of the model artifact and to validate it using end to end tests before field use.

9.3.2 Computing Environment Constraints

Rig site compute limitations can constrain transformer inference and ensemble methods. Where inference is required at short measure, model compression or distillation may be needed (Moslemi et al. 2024). Deployment choices should be aligned with the decision measure, for example slower, higher fidelity forecasts for planning tasks, and lighter models for second scale monitoring tasks.

9.3.3 Human Factors and Actionability

A final limitation is that the case studies focus on predictive accuracy and reconstruction quality, but operational adoption depends on actionability. Forecasts and anomaly flags must map to recommended checks or parameter adjustments, and these recommendations must be expressed with clear confidence and known failure modes. In practical deployments, the workflow should specify what occurs when the model confidence is low, including fallback logic and escalation to human review. This is particularly important for denoising and diagnostics, where false positives can distract crews and false negatives can delay intervention.

9.3.4 Future Directions

Future work that extends the results in this volume should prioritize three directions. First, expand evaluation across wells, tool strings, and acquisition setups to quantify external validity. Second, strengthen task aligned metrics for telemetry and diagnostics by incorporating downstream success criteria, not only reconstruction error. Third, formalize distribution shift monitoring and uncertainty calibration checks so that reliability can be assessed continuously during operations, not only during offline evaluation.

References

Altindal, M. C., Nivlet, P., Tabib, M., Rasheed, A., Kristiansen, T. G., & Khosravanian, R. (2024). Anomaly detection in multivariate time series of drilling data. Geoenergy Science and Engineering, 237, 212778.

Aoun, M. A. B., & Madarász, T. (2022). Applying machine learning to predict the rate of penetration for geothermal drilling located in the Utah FORGE site. Energies, 15(12), 4288.

Du, S., Huang, C., Ma, X., & Fan, H. (2024). A review of data-driven intelligent monitoring for geological drilling processes. Processes, 12(11), 2478.

Hermawan Manuaba, I. B. G., Aljishi, M., Van Steene, M., & Dolan, J. (2024). Logging-while-drilling laterolog vs. electromagnetic propagation measurements: Which is telling the true resistivity? SPE Journal, 29(8), 4000–4013.

Iglesias, G., Talavera, E., González-Prieto, Á., Mozo, A., & Gómez-Canaval, S. (2023). Data augmentation techniques in time series domain: A survey and taxonomy. Neural Computing and Applications, 35(14), 10123–10145.

Li, Lixin, et al. (2021). A Review of the Research and Development of High-Frequency Measurement Technologies Used for Nonlinear Dynamics of Drillstring. Shock and Vibration 2021, 1, 8821986.

Moslemi, Amir, et al. (2024). A survey on knowledge distillation: Recent advancements. Machine Learning with Applications, 18, 100605.

Mwachaka, Saleh M., Aiping Wu, and Qingqing Fu. (2019). A review of mud pulse telemetry signal impairments modeling and suppression methods. Journal of Petroleum Exploration and Production Technology 9(1), 779–792.

Nunes, P., Santos, J., & Rocha, E. (2023). Challenges in predictive maintenance–A review. CIRP journal of manufacturing science and technology, 40, 53–67.

Okoh, C., Roy, R., Mehnen, J., & Redding, L. (2014). Overview of remaining useful life prediction techniques in through-life engineering services. Procedia CIRP, 16, 158–163.

Polo, Felipe Maia, et al. (2023). A unified framework for dataset shift diagnostics. Information Sciences, 649, 119612.

The manufacturer's authorised representative in the EU is Springer Nature Customer Service Centre GmbH, Europaplatz 3, 69115 Heidelberg, Germany. If you have any concerns regarding our products, please contact ProductSafety@springernature.com

Printed and bound by CPI Group (UK) Ltd, Croydon, CR0 4YY
05/10/2026
02228059-0008